Mathematics
Worksheets
Don't Grow Dendrites

Other Corwin Press Books by Marcia L. Tate

Shouting Won't Grow Dendrites: A Multimedia Kit for Professional Development (2008)

Shouting Won't Grow Dendrites: 20 Techniques for Managing a Brain-Compatible Classroom (2006)

Worksheets Don't Grow Dendrites: A Multimedia Kit for Professional Development (2006)

Reading and Language Arts Worksheets Don't Grow Dendrites: 20 Literacy Strategies That Engage the Brain (2005)

"Sit and Get" Won't Grow Dendrites: 20 Professional Learning Strategies That Engage the Adult Brain (2004)

Worksheets Don't Grow Dendrites: 20 Instructional Strategies That Engage the Brain (2003)

Mathematics Worksheets Don't Grow Dendrites

20

Numeracy Strategies That Engage the Brain, PreK-8

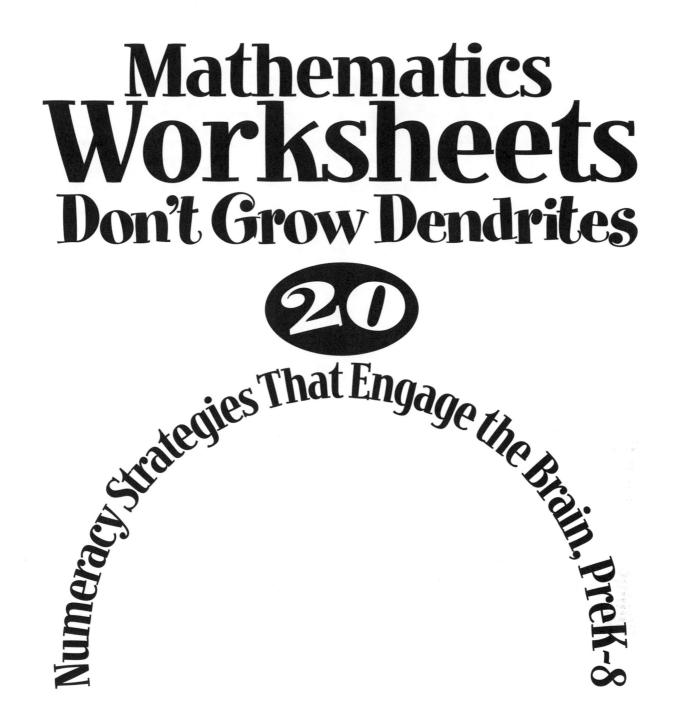

Marcia L. Tate

CORWIN PRESS
A SAGE Company

Illustrations by Robert Greisen.

For information:

Corwin Press
A SAGE Company
2455 Teller Road
Thousand Oaks, California 91320
www.corwinpress.com

SAGE Pvt. Ltd.
B 1/I 1 Mohan Cooperative Industrial Area
Mathura Road, New Delhi 110 044
India

SAGE Ltd.
1 Oliver's Yard
55 City Road
London EC1Y 1SP
United Kingdom

SAGE Asia-Pacific Pte. Ltd.
33 Pekin Street #02-01
Far East Square
Singapore 048763

Printed in the United States of America.

Library of Congress Cataloging-in-Publication Data

Tate, Marcia L.
Mathematics worksheets don't grow dendrites: 20 numeracy strategies that engage the brain, PreK–8/Marcia L. Tate.
   p. cm.
Includes bibliographical references and index.
ISBN 978-1-4129-5332-0 (cloth)
ISBN 978-1-4129-5333-7 (pbk.)
   1. Mathematics—Study and teaching (Elementary)—Activity programs.
2. Mathematics—Study and teaching (Middle school)—Activity programs.
3. Mathematics—Study and teaching (Elementary)—United States. 4. Mathematics—Study and teaching (Middle school)—United States. 5. Effective teaching. I. Title.

QA135.6.T34 2009
372.7—dc22   2008017908

This book is printed on acid-free paper.

08   09   10   11   12   10   9   8   7   6   5   4   3   2   1

Acquisitions Editors:	Allyson Sharp, Carol Chambers Collins
Editorial Assistants:	David Gray, Brett Ory
Production Editor:	Veronica Stapleton
Copy Editor:	Jovey Stewart
Typesetter:	C&M Digitals (P) Ltd.
Proofreader:	Tracy Marcynzsyn
Indexer:	Sheila Bodell
Cover Designer:	Lisa Miller
Graphic Designer:	Monique Hahn

Contents

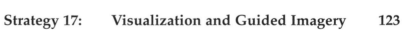

Introduction

During more than 30 years in education, I have learned a great deal about what teachers should and should not do when it comes to managing their classrooms. I have actually seen both of the following scenarios several times:

SCENARIO I

I hate my math class!" Brian could be overheard saying to his friend Paul as they walked down the hall of Kennedy Middle School. "It doesn't even matter whether I'm in class or not! I never understand what Mrs. Stabler is teaching and if I don't understand what we're doing the first time she explains it, she won't re-explain. She gives us math to do for homework but since I don't know what I'm doing in class, I sure can't begin to answer the homework, so I get them all wrong. This leads to another 'F' in her grade book."

Today the topic for the lesson is Data Analysis and Probability. Mrs. Stabler is required to teach students to figure out the *mean*, *mode*, and *median* of sets of data. As Paul reluctantly files into class, he is asked to take his seat and open his math book to page 35. He and other students comply. Mrs. Stabler explains to students how to find the *mode* of a set of data and tells them that it is the most frequently occurring score. She then immediately begins assigning exercises to students. She calls on Paul to solve the fourth exercise. Paul answers correctly and is surprised that he finally gets one answer right. This practice continues until all of the exercises on page 35 have been answered by Paul's peers.

Mrs. Stabler then proceeds to explain how to find the *mean* by working several exercises on the board. She tells students that the word average is another name for the word *mean*. This leads to the next set of practice exercises that students are asked to solve on their papers. When two students begin to talk while they are supposed to be working independently, they are promptly and harshly reprimanded.

Although several students have questions, time is of the essence so Mrs. Stabler must move on to the last concept of the day, the *median*. The routine is the same. Work several exercises on the board, assign 10 or 15 problems as seat work for the class, have students solve the exercises independently, and then go over them orally.

The bell rings and Paul and his peers proceed to their next class. It's just another day in "paradise" or, should I say, math class in middle school.

SCENARIO II

Today the focus in math class is Data Analysis and Probability. Mr. Rutledge at McNair Middle School is also required to teach students to figure out the *mean*, *mode*, and *median* of sets of data. As students file into his math class, there is an air of exuberance. You see, yesterday, Mr. Rutledge told students that five of them could volunteer to bring a CD containing their favorite song to class. Provided that the lyrics of the song are acceptable, Mr. Rutledge will play each song for the class today. While almost all students volunteered to be the lucky ones to bring their music, only five students were selected. Today those five students have their songs, and Mr. Rutledge has the CD player. He will play each song and ask every student in class to rate all five songs on a scale of one to 10, where 10 is the highest rating and one is the lowest. Once the songs have been played, Mr. Rutledge now has five sets of data from which the class can derive the mean, mode, and median.

Need, *novelty*, *meaning*, and *emotion* are four ways the brain can be hooked into a lesson. (Refer to Resource B: Brain-Compatible Lesson Design in this book for more in-depth information on ways to hook students.) If a lesson has even one of these, it stands a good chance of being remembered. The activity that Mr. Rutledge has planned provides not one, but all four hooks that he can use to interest students in learning about this math concept.

Students are anxious to find out which song has the highest rating and therefore have a purpose for learning to figure out the mean, mode, and median of the song data. They have a *need* to know. The lesson is also being taught in a new and different way. This makes the instruction *novel*. When sixth graders are listening to their own music, they are certainly connected to the real world, which ensures that the lesson has *meaning*. Finally, because students are excited about the lesson, enthusiasm and *emotion* are present, which correlates with long-term retention.

Once the students are "hooked" into the instruction and have already listened to their music, Mr. Rutledge uses three additional brain-compatible strategies during the lesson itself. He pulls some of the data

gathered from the song ratings and *visually* shows students how to figure out the mode, the most frequently occurring score. He then asks for choral responses from the class and calls on volunteers and non-volunteers as they *discuss* the procedure for finding the median. Students then work with a partner to find the mean for each song, *reciprocally teaching* the procedures as the conversation progresses. They finally reach conclusions as to the class' favorite song.

BRAIN-COMPATIBLE INSTRUCTION

The above scenarios describe two math classes, both teaching the same content, but using very different methodology. When you consider which teacher stands the best chance of getting the concept of mean, mode, and median into the long-term memory of students' brains, no doubt you would consider the latter class. Why?

One learns to do by doing.

—Aristotle

Tell me, I forget.
Show me, I remember.
Involve me, I understand!

—Old Chinese Proverb

Thousands of years of history support one major concept. When students are engaged in firsthand experiences with content, they stand a better chance of learning what they need to know. The mathematics curriculum is no exception. Yet the increased emphasis on *high-stakes testing* encourages teachers to spend a great deal of time on low-level concepts that can be easily measured by paper and pencil tests, thus turning the classroom into a boring, unmotivated wasteland. When students simply memorize a mathematics formula, they perceive math as a boring, meaningless mix of rules given by the teacher and memorized by the students (Posamentier & Hauptman, 2006). Unfortunately, this type of teaching can also occur to the exclusion of those teaching strategies that foster stimulating lessons, retention of information, and increased academic achievement.

Learning-style theorists (Gardner, 1983; McCarthy, 1990; and Sternberg & Grigorenko, 2000) and educational consultants (Jensen, 2001; Sousa, 2006, 2007; Tate, 2003; Wolfe, 2001) have figured out that some instructional strategies simply work better for learning than others since, by their very nature, they result in long-term retention of information. These strategies have been summarized and applied in the following four bestsellers and will be outlined in the paragraphs that follow: *Worksheets Don't Grow Dendrites: 20 Instructional Strategies That Engage the Brain* (Tate, 2003); *"Sit and Get" Won't Grow Dendrites: 20 Professional Learning Strategies That Engage the Adult Brain* (Tate, 2004); *Reading and Language Arts Worksheets Don't Grow Dendrites: 20 Literacy Strategies That Engage the Brain* (Tate, 2005); and *Shouting Won't Grow Dendrites: 20 Techniques for Managing a Brain-Compatible Classroom* (Tate, 2006). These strategies will form the foundation of the remainder of this text. In addition the activities contained in this book will demonstrate the use of the 20 strategies for teaching that reflects the National Council of Teachers of Mathematics' focal points (NCTM, 2006). (See list of Focal Points below.)

During the 1990s, then President Herbert Walker Bush declared that this was the *decade of the brain.* Today, Paul Allen, cofounder of Microsoft, and others are investing millions in the continued study of this wondrous structure. Teachers should be the first to avail themselves of this information since they are teaching the brains of students each and every day. In fact, I tell teachers that the next time they complete a resume, they need to include that they are not only *Teachers,* but also *Dendrite* (brain cell) *Growers!*

For the past 12 years, I have been studying this absolutely awesome structure called the brain. Through my extensive reading and participation in workshops and courses with experts on the topic, as well as my observations of best classroom practices, I have synthesized the instructional strategies most effective in teaching the brain into 20 methods of delivering instruction. Whether you are also studying the latest brain research (Jensen, 2001; Sousa, 2006, 2007; Wolfe, 2001) or learning style theories (Sternberg & Grigorenko, 2000) you will come up with similar conclusions. If you want students to retain content long-term, there are 20 ways to deliver instruction. Teachers who use these strategies with all students not only have classrooms where students excel academically and where behavior problems are reduced, but also where teaching and learning become fun since students who laugh together, learn together. The 20 strategies are as follows:

- Brainstorming and discussion
- Drawing and artwork
- Field trips
- Games
- Graphic organizers, semantic maps, and word webs
- Humor
- Manipulatives, experiments, labs, and models
- Metaphors, analogies, and similes
- Mnemonic devices
- Movement
- Music, rhythm, rhyme, and rap
- Project-based and problem-based instruction
- Reciprocal teaching and cooperative learning
- Role plays, drama, pantomimes, and charades
- Storytelling
- Technology
- Visualization and guided imagery
- Visuals
- Work study and apprenticeships
- Writing and journals

Refer to Figure I on page xix for a correlation of these 20 strategies to Howard Gardner's (1983) *Theory of Multiple Intelligences* as well as to the four major modalities—visual, auditory, kinesthetic, and tactile. Each lesson that incorporates multiple modalities stands a better chance of being remembered by students long after the teacher-made, criterion-referenced, or standardized tests are over. After all, isn't that what matters—long-term retention?

By answering the question words, *who, what, when, where,* and *how,* the book you are about to read attempts to accomplish five major objectives:

1. Delineate each of the 20 brain-compatible strategies that take advantage of how the brain learns best in the following 20 chapters.

2. Review the research regarding the 20 strategies as they relate to teaching the major focal points of mathematics.

3. Supply over 200 examples of the application of the 20 strategies in teaching each of the focal points at a variety of grade levels.

4. Provide time and space at the end of each chapter for the reader to reflect on the application of the strategies for teaching their own curricular objectives.

5. Demonstrate how to plan and deliver unforgettable lessons by asking the five questions on the lesson plan format contained in Resource B.

The brain-compatible activities in each chapter are only samples of lessons that can be produced when the strategies are incorporated from prekindergarten to pre-calculus. They are intended only to get the reader's brain cells going as they think up a multitude of additional ways to deliver brain-compatible mathematics instruction to their students. While specific grade levels are delineated for the use of the strategies, many can be adapted to other grade levels or used with remedial or advanced students at the same grade level.

When you really examine the list of 20, you will find that they are used most frequently in the lower elementary grades. As the strategies begin to disappear from the repertoire of teachers, so diminishes students' academic achievement, grades, confidence, self-esteem, and love for school. The challenge is becoming so severe that the cover story of *Time* magazine, dated April 7, 2006, was titled "Dropout Nation." It appears that 30% of high school students in the United States are not graduating. In many major inner cities, the number can be as high as 50% to 60%. If a business was losing 30% to 50% of its clients per year, how long would it remain in business? The answer, of course, is *not very long*.

There are a variety of reasons for the aforementioned dilemma, and no one person has all the answers. However, part of the answer lies in the following sign I saw posted on the wall in a teachers' lounge: *If students do not learn the way we teach them, then we must teach them the way they learn.*

There are 20 ways to teach and 20 ways to learn. For example, students in Singapore, who have some of the highest math scores of any country in the world, use two of the strategies extensively as they solve problems in mathematics—visualization and drawing. It is only beneficial and timely for us to think beyond the status quo and begin to teach in ways that are proven to be effective.

NATIONAL COUNCIL OF TEACHERS OF MATHEMATICS FOCAL POINTS

Since early man had to survive in ways that employed quantitative sense (e.g., How many animals are chasing me? How much firewood do I need to last through the night? How many days before the next high tide?), the ability to do basic counting appears to be hardwired into the brain even at birth. However, more complex mathematical processes are not as much of a priority to the brain and necessitate the need for formal or informal instruction (Bender, 2005). This instruction is by no means easy since mathematical thinking involves many different areas of the brain, such as the frontal and parietal lobes, for higher-level thought processes; the visual cortex, for visualizing math problems; and the angular gyrus, Broca's area, and Wernicke's area, which are all involved in the act of reading and comprehending word problems. When students are motivated to learn math, or any subject for that matter, an emotional rationale is introduced which facilitates memory.

The National Council of Teachers of Mathematics was one of the first professional organizations to develop standards. In 1989, they originated outcomes in the following areas that delineate the mathematical concepts and procedures essential for making every student mathematically literate.

Number and Operations
Algebra
Geometry
Measurement
Data Analysis and Probability
Problem Solving
Reasoning and Proof
Communications
Connections
Representation

In 2006, these NCTM standards were consolidated into curriculum focal points. The focal points were designed to provide short and long-term opportunities to improve mathematics instruction by describing the most significant cumulative grade-level skills and concepts. Three focal points per grade level from prekindergarten through Grade 8 have been identified and will be described in the paragraphs that follow. While these focal points do not provide specific approaches for instruction, they are designed to be taught within the context of the following mathematical processes: reasoning and proof, problem solving, connections, communication, and representation. These focal points describe the most significant cumulative grade-level skills and concepts.

Number and Operations

Beginning in prekindergarten and kindergarten, students develop an understanding of whole numbers and how to use those numbers to represent, compare, and order quantities. By Grade 1, they are developing strategies for adding and subtracting those whole numbers, including the use of properties of addition and more sophisticated strategies. Grade 2 students are counting using base-10 numeration systems and concepts of place value while Grade 3 are using representations to comprehend the meanings of multiplication and division of whole numbers. Grade 4 students are recalling multiplication and division facts more rapidly and becoming more fluent with multiplying whole numbers. By Grade 5, this focal point is directly connected to Algebra because students are developing more fluency with procedures for dividing whole numbers and using patterns, models, and relationships for writing and solving simple equations and inequalities. Grades 6 through 8 enable students to develop fluency with multiplication and division of fractions as well as an application of ratio, proportionality, and systems of linear equations.

Algebra

While Algebra is integrated with a focal point at each grade level beginning in Grade 1, by Grade 6 it is a separate focal point. By this time, students are writing, evaluating, and using mathematical expressions and formulas to solve problems.

Geometry

Geometry is a separate focal point at prekindergarten, kindergarten, Grades 1, and 3. It is integrated with other focal points at Grades 5, 7, and 8. Prekindergarten and kindergarten students are identifying two- and three-dimensional shapes in their environment and describing those shapes with appropriate vocabulary. By Grade 1, students are building an understanding of part–whole relationships by composing and decomposing both plane and solid figures. At Grade 3, students are transforming polygons to make other polygons as they describe, analyze, and classify shapes by their angles and sides. Grade 6 students are comparing two-dimensional shapes to three-dimensional shapes and analyzing polyhedral solids by the number of edges, faces, and vertices. Geometry is directly paired with measurement at Grade 8 since students are solving problems by using what they know about distance and angles to analyze situations in two- and three-dimensional space.

Measurement

While measurement is connected to focal points at all grade levels, it is directly addressed at prekindergarten, kindergarten, Grades 2, 4, 5, 7, and 8. Beginning at prekindergarten and kindergarten, students are comparing the measurable attributes of objects and solving problems by ordering and comparing those attributes. By Grade 2, their understanding is more developed to include concepts of linear measurement, such as transitivity and partitioning. At Grade 4, students comprehend that a square is a standard unit of measurement for determining area and that the area of two-dimensional shapes can be determined.

Measurement is integrated with Geometry and Algebra at Grades 5 and 7 as students learn to describe two- and three-dimensional shapes. At these grade levels, they should be able to choose the appropriate units and strategies for estimating and measuring volume, and justifying formulas for the surface areas and volumes of prisms and cylinders when solving real-world problems. By Grade 8, students are applying their reasoning about similar triangles to solve various real-world problems. They are applying the Pythagorean Theorem to analyze polygons and polyhedra and to determine the distances between points in the Cartesian coordinate plane.

Data Analysis and Probability

At Grade 8, the focal points of Number and Operations and Algebra are directly integrated with Data Analysis since, by this time, students are summarizing and comparing data sets by using descriptive statistics as well as posing and answering questions by organizing and displaying data.

NATIONAL MATH ADVISORY PANEL REPORT

In 2006, a National Math Advisory Panel was appointed to ascertain the best research regarding the effective teaching and learning of mathematics. This report titled, *Foundations for Success,* is divided into six recommendations, which are summarized below:

1. The mathematics curriculum needs to be streamlined into a coherent set of topics which are crucial for all students to experience at specified grade levels. These topics should prepare students for the study of algebra and beyond.

2. All students are capable of learning mathematics if conceptual understanding, computational ability, and problem-solving competency are developed simultaneously.

3. Effective teachers must understand math content as well as how students learn, have high expectations for student success, employ a wide range of teaching strategies, and become life-long professional learners.

4. Instruction should be influenced by a combination of high-quality research and the best judgment of professional and experienced teachers.

5. Assessment that informs and guides teacher decision making should be an integral part of instruction and should enhance student learning.

6. Mathematical capacity should be continuously expanded with more research in the areas of technology, teacher preparation, and instructional practices. (NCTM, 2008)

Many of the aforementioned recommendations will be reflected in the brain-based strategies that follow.

The activities outlined in each chapter are designed to be starting points for planning lessons that are intended to be brain-compatible and are in no way meant to be an exhaustive list of possibilities. The advantage of having activities that range from prekindergarten through eighth grade in the same book is that the reader can easily select activities that will meet the needs of students performing below, on, and above grade level and can therefore more easily differentiate instruction. You will also find that an activity designated for a specific grade range can be taken as is or easily adapted to fit the grade level the reader is teaching. Therefore as you peruse this text, examine not only those activities in each strand that are age or grade appropriate, look for ones at other grade levels that can easily meet your needs once you change the conceptual level of the material.

The Reflection and Application page at the end of each chapter enables readers to apply the activities read to their own students or to enter activities that they have created. The Lesson Design section (see Resource B) helps the reader synthesize the process of planning unforgettable lessons by asking the following five pertinent questions:

1. What will you be teaching?

2. How will you know students have learned the content?

3. How will you gain and maintain students' attention?

4. How will you divide and teach the content to engage students' brains?

5. Which of the 20 brain-compatible strategies will you use to deliver content?

Enjoy applying the brain-compatible strategies to the NCTM focal points and making mathematicians out of students who never thought it possible!

Brain-Compatible Strategies	Multiple Intelligences	VAKT
Brainstorming and discussion	Verbal-linguistic	Auditory
Drawing and artwork	Spatial	Kinesthetic/tactile
Field trips	Naturalist	Kinesthetic/tactile
Games	Interpersonal	Kinesthetic/tactile
Graphic organizers, semantic maps, and word webs	Logical-mathematical/ spatial	Visual/tactile
Humor	Verbal-linguistic	Auditory
Manipulatives, experiments, labs, and models	Logical-mathematical	Tactile
Metaphors, analogies, and similes	Spatial	Visual/auditory
Mnemonic devices	Musical-rhythmic	Visual/auditory
Movement	Bodily-kinesthetic	Kinesthetic
Music, rhythm, rhyme, and rap	Musical-rhythmic	Auditory
Project-based and problem-based instruction	Logical-mathematical	Visual/tactile
Reciprocal teaching and cooperative learning	Verbal-linguistic	Auditory
Role plays, drama, pantomimes, charades	Bodily-kinesthetic	Kinesthetic
Storytelling	Verbal-linguistic	Auditory
Technology	Spatial	Visual/tactile
Visualization and guided imagery	Spatial	Visual
Visuals	Spatial	Visual
Work study and apprenticeships	Interpersonal	Kinesthetic
Writing and journals	Intrapersonal	Visual/tactile

Figure I   Comparison of Brain-Compatible Instructional Strategies to Learning Theory

Acknowledgments

Of all of my literary attempts, this has been by far the most challenging. This is probably due to the fact that I am a reading specialist and by no means an authority in the area of mathematics. However, as I have observed in numerous math classes over my 35-year tenure, I have witnessed the best math teachers incorporating brain-compatible strategies into their lessons, even if they had never had a course on brain-compatible instruction. They just instinctively seemed to know what was best for students! I have had to elicit the help of some of these teachers and the activities of 17 of them have been included in this book. I thank them for the submission of their ideas which have only improved the quality of this text. Their names appear under the activities submitted.

A special debt of gratitude goes to Thomasenia Lott Adams, PhD, mathematics education professor at the University of Florida and to Allyson Sharp, my Corwin Press editor, for their patience, guidance, and recommendations for improving my original manuscript.

Once again, I acknowledge my family and the support of my husband, Tyrone; my mother, Eurica; and my three wonderful children, Jennifer, Jessica, and Christopher. I also appreciate the encouragement of my son-in-law Lex, and daughter-in-law, Amanda. Now that I have completed another book, we can play *Scrabble* more often.

Since my last book, I have also become a grandmother! If I had known how wonderful it would be to have grandchildren, I would have skipped having children and gone straight to the grandchildren. Only kidding! I also trust that this and other books about brain-compatible instruction equip the teachers of my grandchildren with the knowledge and skill that will enable them to maximize their potential and the potential of all students.

Thanks also to my associates and executive assistants Carol and Sadira for what you do daily for our company, Developing Minds Inc.

About the Author

 **Marcia L. Tate, EdD,** is the former executive director of professional development for the DeKalb County School System, Decatur, Georgia. During her 30-year career with the district, she has been a classroom teacher, reading specialist, language arts coordinator, and staff development executive director. She received the Distinguished Staff Developer Award for the State of Georgia, and her department was chosen to receive the Exemplary Program Award for the state.

Marcia is currently an educational consultant and has taught more than 175,000 administrators, teachers, parents, and business and community leaders throughout the world. She is the author of the following four bestsellers: *Worksheets Don't Grow Dendrites: 20 Instructional Strategies That Engage the Brain; "Sit & Get" Won't Grow Dendrites: 20 Professional Learning Strategies That Engage the Adult Brain; Reading and Language Arts Worksheets Don't Grow Dendrites: 20 Literacy Strategies That Engage the Brain;* and *Shouting Won't Grow Dendrites: 20 Techniques for Managing a Brain-compatible Classroom.* Participants in her workshops refer to them as "the best ones they have ever experienced" since Marcia uses the 20 strategies outlined in her books to actively engage her audiences.

Marcia received her bachelor's degree in psychology and elementary education from Spelman College in Atlanta, Georgia. She earned her master's degree in remedial reading from the University of Michigan, her specialist degree in educational leadership from Georgia State University, and her doctorate in educational leadership from Clark Atlanta University. Spelman College awarded her the Apple Award for excellence in the field of education.

Marcia is married to Tyrone Tate and is the proud mother of three children: Jennifer, Jessica, and Christopher and the doting grandmother of two grandchildren, Christian and Aidan. Marcia can be contacted by calling her company at (770) 918-5039 or by e-mail: marciata@bellsouth.net. Visit her Web site at www.developingmindsinc.com.

Brainstorming and Discussion

WHAT: DEFINING THE STRATEGY

The answer is 32. What is the question?

How did you get the answer to the third problem?

What was the first thing you thought about as you began to tackle this equation?

What would have happened if we had divided instead of multiplied?

According to research about how the brain behaves, the person in the classroom who is doing the most talking about the content is growing the most dendrites or brain cells. My observations in many classrooms have led me to believe that this person is the teacher. A large number of teachers grow dendrites daily since they are the only ones talking about math. The problem is that students are not a part of the conversation. I recently observed in a classroom where a teacher was solving the calculus homework problems on the board without ever asking students a single question or engaging them in any discussion regarding the thinking processes accompanying their individual answers.

Discussion not only provides the brain with a supply of oxygen that keeps it in a more alert state during the lesson, it also facilitates memory. The open-ended questions listed above provide students with opportunities to brainstorm ideas and engage in lively classroom conversation. Whether you are instructing the whole class or having students work in small, flexible groups, learning is facilitated when instructors ask open-ended questions and acknowledge and encourage a variety of ideas as students engage in interactive discourse.

WHY: THEORETICAL FRAMEWORK

The most widely known technique for stimulating creativity in the brain is probably the act of brainstorming where all ideas are accepted and there is a greater chance of reaching a workable solution (Gregory & Parry, 2006).

Students with special needs benefit when the class works in groups of less than six and the teacher uses directed-response questioning so that students have a chance to think aloud (Jensen, 2005).

Teachers can guide students through very difficult solutions by using a series of well-thought-out questions that address process rather than procedure (Posamentier & Jaye, 2005).

Class discussion can assist students in comprehending the properties of operations, such as the associative and communicative properties (National Council of Teachers of Mathematics, 2000).

Discussion and questioning during whole class or cooperative group learning enable the brain to clarify concepts and hook new information with the information that the brain already knows (Brooks & Brooks, 1993).

When students share their thinking about number combinations in class discussions, other students are able to develop or improve their strategies (National Council of Teachers of Mathematics, 2000).

HOW: INSTRUCTIONAL ACTIVITIES

WHO: Kindergarten–Grade 8
WHEN: Before the lesson
FOCAL POINT(S):   All

• Prior to the teaching of a lesson, have students brainstorm all they know about the particular concept that will be addressed in the lesson. For example, prior to a lesson on fractions, have students brainstorm all the occasions for which fractions are used in everyday living, such as following recipes, dividing pizza into slices, and calculating sale prices. Also engage students in listing all of the questions they have about the topic of fractions.

WHO: Prekindergarten–Grade 8
WHEN: During the lesson
FOCAL POINT(S):   All

• When asking math questions in class or creating teacher-made math tests, provide opportunities for all students to be successful by asking both knowledge or short-answer questions as well as those that enable students to use their mathematical reasoning and critical and creative-thinking skills. Refer to Figure II on pages 5–7 to ensure that students have opportunities to answer questions at all levels of Bloom's Taxonomy, particularly those above the *Knowledge* level.

WHO: Grade 3–Grade 8
WHEN: During or after the lesson
FOCAL POINT(S):   All

• During cooperative group discussions or as students create original questions for math assessments following a unit of study, have them use the question stems in Figure II on pages 5–7. These stems will help to ensure that questions are created that represent all levels of Bloom's Taxonomy.

WHO: Prekindergarten–Grade 2
WHEN: During the lesson
FOCAL POINT(S):   Number and Operations

• When teaching the whole class, have students brainstorm answers to discussion questions, such as those that require them to use their number sense. Questions should be similar to the following:

1. How many boys (girls) are in our class?
2. If we subtract the number of boys from the number of girls, how many students are left?
3. Show me five fingers. If we take away two fingers, how many fingers are left?

Sentence starters similar to the ones listed above are particularly effective for English language learners since they enable these students to take an active part in the discussion:

• I realize that . . .
• I agree with _____that _____.
• I would like to add to _____'s idea.
• I don't understand what _____ meant when she said . . . (Coggins, Kravin, Coates, & Carrol, 2007).

WHO: Kindergarten–Grade 8
WHEN: During the lesson
FOCAL POINT(S):   All

• Use the Think, Pair, Share technique with students. Pose a question or discussion topic to the class. Have them think of an individual answer. Then have them pair with a peer and share their answer. Then call on both volunteers and non-volunteers to respond to the entire class.

WHO: Grade 3–Grade 5
WHEN: During the lesson
FOCAL POINT(S):   Data Analysis

• Have students discuss the numbers of brothers and sisters, if any, they have. As students give you the numbers, begin listing them in columns on the board. Discuss that some students have the same number of siblings and others have different numbers; some have more and some have less, which introduces the concept of range (difference between the greatest and the least) and the concept of mode (most common number of siblings). Other measures of central tendency that can be introduced in the discussion are mean and median.

WHO: Prekindergarten–Grade 8
WHEN: During the lesson
FOCAL POINT(S):   All

• Engage students in a class discussion that would help them to form generalizations. Ask questions such as *How would you describe this pattern? How can this same pattern be repeated?* (National Council of Teachers of Mathematics, 2000).

WHO: Grade 1–Grade 8
WHEN: After the lesson
FOCAL POINT(S):   All

• Following the completion of a homework assignment, have students compare their answers with one another. If students' answers differ, have them use discussion to defend their answers to their partners and reach consensus as to which answers are acceptable and why.

WHO: Prekindergarten–Grade 8
WHEN: During the lesson
FOCAL POINT(S):   All

• Ask students a question to which there is more than one appropriate answer. Have students brainstorm as many ideas as possible in a designated time period using the **DOVE** guidelines:
    o *D*efer judgment
    o *O*ne idea at a time
    o *V*ariety of ideas
    o *E*nergy on task (Tate, 2003)

WHO: Prekindergarten–Grade 8
WHEN: During the lesson
FOCAL POINT(S):   All

• When asking a discussion question, wait a minimum of five to seven seconds to allow students' brains the opportunity to reason out the answer. If after a five-second minimum the student does not respond, either rephrase the question, provide additional information, give a clue, or provide the student with question structures or frames such as the following:
    o Why is _____ different from _____?
    o How is this answer similar to the previous answer?
    o What is another way to say it?

WHO: Grade 2–Grade 8
WHEN: After the lesson
FOCAL POINT(S): All

- After engaging in problem-solving experiences, involve students in discussion about solutions or lack thereof:
    - Is there another way to solve the problem?
    - Could there be other solutions to the problem?
    - What happens if we change (some variable) in the problem?
    - Explain your answer.
    - Create a similar problem—one that is solved in the same way or one that has a similar answer.

WHO: Grade 6–Grade 8
WHEN: During the lesson
FOCAL POINT(S): Algebra

- Replicate or model previous students' work (not even your own students) that has errors. Display this work for the class; then engage the class in brainstorming what errors are present and what may have led to the learner making such errors. Be certain the names of students remain anonymous.

Model Questions and Key Words to Use in Developing Questions

I. **Knowledge** (Eliciting factual answers, testing recall and recognition)

Who	Where	Describe	Which one
What	How	Define	What is the best one
Why	How much	Match	Choose
When	What does it mean	Select	Omit

II. **Comprehension** (Translating, interpreting, and extrapolating)

State in your own words	Classify	Which are facts, opinions
What does this mean	Judge	Is this the same as
Give an example	Infer	Select the best definition
Condense this paragraph	Show	What would happen if

State in one word

What part doesn't fit

What restrictions would you add

What exceptions are there

Which is more probable

What are they saying

What seems to be

What seems likely

Indicate

Tell

Translate

Outline

Summarize

Select

Match

Explain

Represent

Demonstrate

Explain what is happening

Explain what is meant

Read the graph, table

This represents

Is it valid that

Which statements support the main idea

Sing this song

Show in a graph, table

III. Application (to situations that are new, unfamiliar, or have a new slant for students)

Predict what would happen if

Choose the best statements that apply

Select

Judge the effects

What would result

Explain

Identify the results of

Tell what would happen

Tell how, when, where, why

Tell how much change there would be

IV. Analysis (Breaking down into parts, forms)

Distinguish

Identify

What assumptions

What motive is there

What conclusions

Make a distinction

What is the premise

What ideas apply, not apply

Implicit in the statement is the idea of

What is the function of

What is fact, opinion

What statement is relevant, extraneous to, related to, not applicable

What does the author believe, assume

State the point of view of

What ideas justify the conclusion

The least essential statements are

What is the theme, main idea, subordinate idea

What inconsistencies, fallacies

What literacy form is used

What persuasive technique

What relationship between

V. **Synthesis** (Combining elements into a pattern not clearly there before)

Write (according to the following limitations)
Create
Tell
Make
Do
Dance
Choose

How would you test
Propose an alternative
Solve the following
Plan
Design

Make up
Compose
Formulate a theory
How else would you
State a rule
Develop

VI. **Evaluation** (according to some set of criteria, and state why)

Appraise
Judge
Criticize
Defend
Compare

What fallacies, consistencies, inconsistences appear
Which is more important, moral, better, logical, valid, appropriate, inappropriate
Find the errors

Figure II   Model Questions and Key Words to Use in Developing Questions

(TESA Program at the Los Angeles County Office of Education, Phone: 1-800-566-6651. Based on *Bloom's Taxonomy*, developed and expanded by John Maynard, Pomona, CA)

REFLECTION AND APPLICATION

How will I incorporate *brainstorming* and *discussion* into mathematics instruction with my students?

Concept _____

Activity _____

_____

_____

Concept _____

Activity _____

_____

_____

Concept _____

Activity _____

_____

_____

Concept _____

Activity _____

_____

_____

Concept _____

Activity _____

_____

_____

STRATEGY 2

Drawing and Artwork

WHAT: DEFINING THE STRATEGY

I would bet that you have several students in your classroom who are excellent artists. The majority of those students may be male. Many boys' brains have natural artistic ability and while you are teaching, they just may be drawing. Their artwork is probably totally unrelated to the lesson you are teaching. Take heart! The content area of mathematics naturally lends itself to the strategy of drawing. In fact, along with visualization, drawing is one of the two major strategies students in Singapore use when solving math problems (Prystay, 2004), and these students have some of the highest math scores in the world.

Use the creativity in students' brains to your advantage. Think of all the math concepts that can be drawn and encourage future professional artists or architects to use their natural talents to facilitate their understanding of your content. Having students draw pictures to represent the addends of an addition problem or a series of geometric shapes to comprehend parallelograms or vertices is one of the best ways to facilitate conceptual understanding. Having them sketch out each step in a word problem also helps them comprehend the problem itself.

WHY: THEORETICAL FRAMEWORK

When useful, teachers should encourage students to draw pictures which can help them gain more insight by representing abstract concepts graphically (Posamentier & Jaye, 2005).

Math books in Singapore teach students to draw models in an effort to visualize math problems prior to solving them (Prystay, 2004).

Different areas of the brain, including the amygdala and the thalamus, are activated when people are involved in art activities (Jensen, 2001).

When examining 1999 and 2000 test results, students who were enrolled in art design, art appreciation, and studio art achieved scores 47 points higher on mathematics college entrance exams than those students who were not a part of such programs (College Board, 2000).

The thinking that occurs when students are involved in art precedes improvements in the thinking that occurs in other areas of the curriculum (Dewey, 1934).

HOW: INSTRUCTIONAL ACTIVITIES

WHO: Prekindergarten–Grade 2
WHEN: During the lesson
FOCAL POINT(S):   Number and Operations

• Have students sort, classify, and order objects by drawing and coloring them. For example, students could draw 10 objects and then arrange them from smallest to largest or point out which one is in first, fourth, or last place.

WHO: Prekindergarten–Grade 2
WHEN: After the lesson
FOCAL POINT(S):   Number and Operations

• Following a lesson on patterns that you provide, students can come up with their own examples of both repeating and growing patterns and draw pictures that represent both types of patterns.

WHO: Kindergarten–Grade 2
WHEN: During the lesson
FOCAL POINT(S):   Number and Operations

• Give each student a small bag of M&M's and a paper towel. Have them open the bag, lay the candies on the paper towel and count the number of M&M's. To introduce the concept of sets, have them sort the candies by color and count the number of M&M's in each set (color). Have students then draw a bar graph that represents the number of candies of each color.

WHO: Kindergarten–Grade 2
WHEN: During the lesson
FOCAL POINT(S):   Geometry

• To demonstrate the concept of symmetry, have students fold a piece of paper equally in half according to a *hamburger fold* (width-wise) or a *hot dog fold* (length-wise). Have them draw and color a picture that is exactly the same on one side of the fold as it is on the other side of the fold.

WHO: Kindergarten–Grade 2
WHEN: After the lesson
FOCAL POINT(S): Measurement

Gallon man—Have students draw or make from construction paper the following Gallon Man as you teach the relationships between the concepts of cups, pints, quarts, gallons, and so forth.

Gallon Man

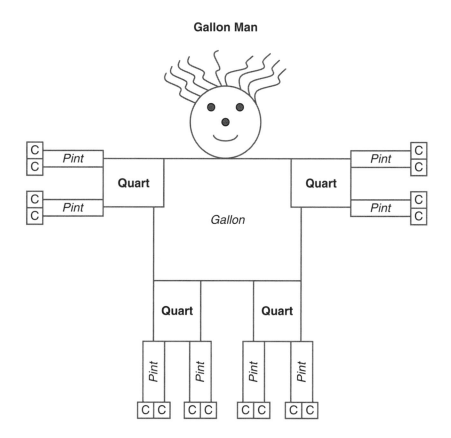

WHO: Grade 3–Grade 5
WHEN: After the lesson
FOCAL POINT(S): Geometry

• Following a lesson, have students draw two-dimensional representations of three-dimensional objects previously taught.

WHO: Grade 3–Grade 5
WHEN: After the lesson
FOCAL POINT(S): Geometry

• Have students draw mathematical terms that could include but are not limited to the following: perpendicular lines, parallel lines, isosceles triangle, right triangle, rhombus, radius, and chord.

WHO: Grade 1–Grade 8
WHEN: During the lesson
FOCAL POINT(S):   Algebra

- Draw four figures on the board that have an obvious pattern and have students draw the four figures on their paper. Then have them predict what the next figure in the series will look like. Have them draw the next four figures that will continue the given pattern. Have them work with a partner to predict what the 10th, 50th, or 100th pattern will look like. Have upper-grade students predict what the nth figure will be.

WHO: Grade 6–Grade 8
WHEN: During or after the lesson
FOCAL POINT(S):   Number and Operations

- Have students draw a capital **S, E,** and **I** as shown below. Tell them that to remember the types of triangles, they need to consider the following:

S Scalene—no equal sides (The S has no equal sides.)

E Equilateral—all three equal sides (The E has three sides.)

I Isoceles—two equal sides (The I has two sides.)

(Shelly Feldman, Grade 3, Noble Avenue Elementary School, North Hills, CA)

WHO: Prekindergarten–Grade 8
WHEN: During the lesson
FOCAL POINT(S):   All

- To assist students in determining the necessary operations in a math word problem, have them read the word problem and then draw a series of pictures that would illustrate what is actually happening in the problem. Have them use the pictures when writing the numerical symbols for the word problem. An example follows:

There are people and dogs in the backyard. All together, they have a combined total of 20 legs. How many people and how many dogs are there in the backyard?

Have students draw people and dogs with the proper number of legs until they figure out the combinations of people and dogs in the yard. These might include: four dogs/two people; one dog/eight people; two dogs/six people; three dogs/four people; four dogs/two people, and so forth.

WHO: Grade 6–Grade 8
WHEN: During or after the lesson
FOCAL POINT(S):   Geometry

- Have students draw histograms, box plots, stem and leaf plots, and scatter plots to depict data when analyzing problems.

WHO: Grade 6–Grade 8
WHEN: During the lesson
FOCAL POINT(S): Probability

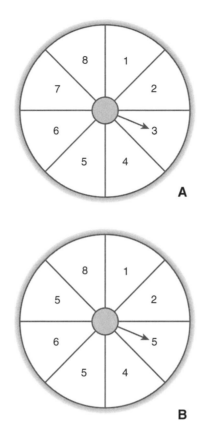

- Draw the following two spinners (A and B) on the board and have students draw them on their papers. Have them determine the probability of spinning the number 5 on Spinner A and on Spinner B. Discuss the differences when determining probability.

WHO: Grade 7
WHEN: During the lesson
FOCAL POINT(S): Geometry

- Have students draw geometric figures, such as a rectangle that measures 4 centimeters by 6 centimeters. Have them label it like the figure below. This and other figures will help them with formulas of measurement.

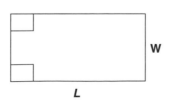

WHO: Grade 6–Grade 8
WHEN: During the lesson
FOCAL POINT(S):   Geometry

• Have students draw a simple map that shows the route they would take in going from their home to school. Students should attempt to draw the map to scale. For example, one mile in reality could equal one inch on their paper.

WHO: Grade 6–Grade 8
WHEN: During the lesson
FOCAL POINT(S):   Geometry

• Have students create tessellations to apply their understandings of symmetry and transformations. A tessellation is a pattern of shapes that is repeated over and over and covers a specific area. The repeated shapes must fit together with no overlaps or gaps.
Use students' tessellations to teach concepts of area, angles, length, congruency, and transformation.

WHO: Grade 6–Grade 8
WHEN: After the lesson
FOCAL POINT(S):   Geometry

• Have students design *matheMADics* drawings which reflect the concept of a geometric term. Some illustrations are as follows:

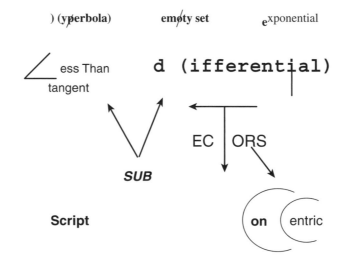

WHO: Grade 6–Grade 8
WHEN: During the lesson
FOCAL POINT(S): Algebra

- Request students to develop visual representation for numerical representations of patterns and vice versa. Consider such patterns as the following:
    - 1, 1, 2, 3, 5, 8, 13, . . . (Fibonacci sequence)
    - 0, 2, 4, 6, 8, 10, . . . (Positive even numbers)
    - 0, 1, 3, 6, 10, . . . (Triangular numbers)

REFLECTION AND APPLICATION

How will I incorporate *drawing* and *artwork* into mathematics instruction with my students?

Concept _____

Activity _____

_____

_____

Concept _____

Activity _____

_____

_____

Concept _____

Activity _____

_____

_____

Concept _____

Activity _____

_____

_____

Concept _____

Activity _____

_____

_____

Field Trips

WHAT: DEFINING THE STRATEGY

Brains have but one purpose—survival in the real world. Is it any wonder that the places that you travel to in the real world are long remembered? This would make the strategy of field trips a memorable one. Visualize a place you visited while you were a student in school. No doubt you still remember the experience. I still remember visiting the Etowa Indian Mounds in Georgia during a unit we were studying on Native Americans. And that was in the second grade. I don't care to tell you how long it has been since I was a student in second grade.

Math concepts are long remembered when they are taught or reinforced in the natural world. However, take the field trip closer to the beginning of the unit rather than at the culmination of it. The concepts in the unit make much more sense when the connections are made during the course of study.

Virtual field trips provide another option. Via the use of technology, students are permitted to *experience* places that could prove inaccessible, inconvenient, or cost prohibitive.

WHY: THEORETICAL FRAMEWORK

Taking students on field trips is one way to incorporate planned movement for learning content into the classroom (Sprenger, 2006b).

Because students need concrete, real-world examples, and need to see, touch, and experience the world, a field trip can be a useful teaching tool prior to starting a teaching unit (Gregory & Parry, 2006).

The classwork of adolescents should carry them into the "dynamic life of their environments" (Brooks, 2002, p. 72).

If students are to link their learning to prior knowledge, they must see the personal connection between what is being taught in the curriculum and their own life experiences (Lieberman & Miller, 2000).

> Concrete experience, not necessarily association, enables the brain to store a great deal of information (Westwater & Wolfe, 2000).
>
> When students get out of the classroom and into the real world, critical thinking skills can improve (Jensen & Dabney, 2000).
>
> Aristotle and Socrates, two of the world's greatest teachers, used field trips thousands of years ago as tools of instruction (Krepel & Duvall, 1981).

HOW: INSTRUCTIONAL ACTIVITIES

WHO: Prekindergarten–Grade 2
WHEN: During the lesson
FOCAL POINT(S):   Number Sense

• Take students on a walk around the school or school grounds. Let them experience their surroundings as you point out objects of interest. Have them stop periodically and count the number of specific objects, such as steps that comprise the stairs, rocks in the yard, leaves on a tree, and so forth.

WHO: Prekindergarten–Grade 5
WHEN: During the lesson
FOCAL POINT(S):   Geometry

• Have students take a look around the room or a walk around the school as well as the community. Ask them to look for patterns in their environment such as in the stars and stripes on the American flag, clothing of classmates, the brick in the school building, or the leaves on the trees. Point out the obvious way that objects, shapes, and colors are patterned in the real world.

WHO: Kindergarten–Grade 2
WHEN: After the lesson
FOCAL POINT(S):   Geometry

• Take students on a walk around the school or school grounds. Have them stop when they see a geometric shape in the real world that you have taught in class, such as the circular bark of a tree, the rectangular wall of the school, or the octagonal shape of a stop sign.

WHO: Grade 2–Grade 5
WHEN: After the lesson
FOCAL POINT(S):   Measurement

• For homework, have students make a list of at least 10 different objects around their home or car that can be measured using a variety of measures or units. Some of the items that could be listed are gas in the car, milk in the refrigerator, water in the bath tub, temperature of the oven, size of the dresser in the bedroom, and so forth. Have students determine which unit is best for measuring each object or the amount the object contains.

WHO: Grade 6–Grade 8
WHEN: After the lesson
FOCAL POINT(S): Geometry

- Following lessons on the concepts of circles, rays, lines, line segments, intersecting and parallel lines, as well as various angles, take the class outside and have them identify the variety of circles, lines, and angles on the sports fields available around the school grounds. Have students measure these fields. Following this field trip, lead students in a discussion on how geometry affects the design of sports fields.

(Francine Rubin, Fourth/Fifth Grade Teacher, Calabash Street Elementary School, Woodland Hills, CA)

WHO: Grade 6–Grade 8
WHEN: During the lesson
FOCAL POINT(S): Geometry

- Have students walk around their community and create math problems from their environment based on what they are discovering. Have them accompany their problems with the photographs, videos, or recordings essential for others to solve the problem. Submit your students' best math problems, along with the accompanying visuals, to the National Math Trails Web site. These problems are then posted to the appropriate grade level for students throughout the United States to solve (Tate, 2003).

WHO: Kindergarten–Grade 8
WHEN: Before the lesson
FOCAL POINT(S): All

- Go online to appropriate Web sites and access virtual field trips that pertain to a math concept being taught.

WHO: Grade 6–Grade 8
WHEN: Before the lesson
FOCAL POINT(S): All

- Take students to observe people at work who use mathematics such as carpenters, landscape planners, and tailors. Small groups of students can be assigned to do this as homework with the help of their parents. Then have them return to the class to share their discoveries about mathematics applications in the real world.

WHO: Grade 6–Grade 8
WHEN: During the lesson
FOCAL POINT(S):   Geometry and Measurement

• Engage students in designing a school or local community mathematics field trip for students in a lower grade. The students can develop an observation or record sheet for the younger students to complete as they engage in the field trip. By thinking about the mathematics younger students will search for on the field trip, the older students are reinforcing their own concepts and understanding of mathematics.

REFLECTION AND APPLICATION

| How will I incorporate *field trips* into mathematics instruction with my students? |

Concept _____

Activity _____

_____

_____

Concept _____

Activity _____

_____

_____

Concept _____

Activity _____

_____

_____

Concept _____

Activity _____

_____

_____

Concept _____

Activity _____

_____

_____

STRATEGY 4

Games

WHAT: DEFINING THE STRATEGY

While prekindergarten children love to play games, it is also one of the 10 activities that keep people living beyond the age of 80, according to the American Association of Retired Persons (AARP). That would lead one to believe that games are beneficial throughout one's life and that elementary, middle, and high school students would benefit from spirited interaction in the pleasurable strategy of game playing. Boys, especially, are naturally motivated when a math review is turned into a friendly competition. Actually, games provide motivation for all students. There is a plethora of game models to choose from, and the imagination is ripe for creating games for children to play and learn at the same time.

Try using the game of *Jeopardy* prior to the testing of a math unit. Place answers to questions or problems on a *Jeopardy* board. The easiest problems would be worth the lowest point value and the most difficult, the highest value. Students then work in three heterogeneous teams to provide the question that accompanies each answer. While the team works together to derive the answer, the team captain provides it orally to the class. Two daily doubles, on which students can wager any or all of their "monies," keep the game interesting, and the bonus round at the end gives added excitement. Students not only practice their conceptual understanding of math but have fun preparing for an upcoming test.

WHY: THEORETICAL FRAMEWORK

Using an event in history, a model, or a game to explore the richness of math are some of the various ways teachers can discover new ways to teach specific math topics (Posamentier & Hauptman, 2006).

> When students perceive their learning environment as positive, endorphins are produced that stimulate the frontal lobes of the brain and give students a feeling of euphoria (Sousa, 2006).
>
> Students not only learn more when playing a game but their participation in class and their motivation for learning math increases (Posamentier & Jaye, 2005).
>
> When students are stressed and perceive their learning environment as negative, cortisol is produced which interferes with the recall of emotional memories (Kuhlmann, Kirschbaum, & Wolf, 2005).
>
> When those students who are going to play the game actually construct it, the game becomes more effective (Wolfe, 2001).
>
> Challenge, feedback, novelty, time, and coherence, all built-in processes fostered by play, enable the brain to mature faster (Jensen, 2001).
>
> Simulations, games, and other types of fun activities gain and maintain the attention of students during a lesson (Burden, 2000).
>
> The need for survival, belonging and love, power, freedom, and fun are the five critical needs that must be satisfied if people are to be effectively motivated (Glasser, 1999).

HOW: INSTRUCTIONAL ACTIVITIES

WHO: Prekindergarten–Grade 8
WHEN: After the lesson
FOCAL POINT(S):   All

- Construct or buy a generic game board, such as Candy Land, with a starting and a finishing point. Have students play in pairs or in small groups. Make game cards to review an objective that has already been taught. For example, cards can include multiplication facts to be recalled, geometric shapes to be identified, or simple algebraic equations to be solved. Students in the group take turns rolling a die or spinning a spinner and then moving separate markers along the game board the same number of spaces rolled. However in order to move the rolled number of spaces, students have to pick the same number of cards and identify the concept on each card. Students move one space for every concept they identify correctly. The first student to get to the end of the game is the winner.

WHO: Prekindergarten–Grade 2
WHEN: After the lesson
FOCAL POINT(S):   Number and Operations

- Have students play any of the games from the appropriate game books listed in Resource A: Have You Read Any Math Lately? starting on page 147 of this book. These would include any of the books from the *Pigs* series by Amy Axelrod.

WHO: Prekindergarten–Grade 2
WHEN: During the lesson
FOCAL POINT(S):   Geometry

• Play the game I Spy by telling students that you are spying a specific object in the room that corresponds to a shape previously studied. The first person to correctly guess the object wins that round. For example, I spy a red circle in the room. That red circle could be the pencil holder on your desk. I spy a white rectangle in the room. The white rectangle could be the dry erase board attached to the wall.

WHO: Grade 1–Grade 3
WHEN: During the lesson
FOCAL POINT(S): Number and Operations

• Have students work in pairs. Give each pair a deck of cards. Have students deal the deck equally between the two of them. Have students hold their half-deck in their hands with the cards face down. Have them turn the top cards up simultaneously and add the value of the two cards together. For example, if one student turns up a seven and another, a three, then the first student to say 10 gets both cards. Jacks are worth 11 points, Queens are worth 12, and Kings are worth 13. Aces can be worth either one or 14 points. The first student to take all the cards or the one who has the most cards when the time is up is named the winner. You may want to pair students with like abilities together.

WHO: Prekindergarten–Grade 2
WHEN: During the lesson
FOCAL POINT(S): Number and Operations

• The simple game of "I'm thinking of a number (insert characteristic)" is beneficial for helping students build number sense and skill with mental comparison and calculation.

WHO: Grade 3–Grade 8
WHEN: During the lesson
FOCAL POINT(S): Number and Operations

• Engage students in the daily mental activity that I have named "Mathnastics," a combination of math and gymnastics. To improve students' ability to do mental math, start every day with a mental math question such as the following:

> Question 1: What is 6 plus 3, minus 4, times 5, divided by 5, plus 2?
> Answer: 7

> Question 2: What is 50, minus 4, divided by 2, minus 11, times 3?
> Answer: 36

The student who is able to give the answer first is the winner. However to give more students an opportunity to get the answer correct, you may want to have all students write the answer down when you say "answer, please." Each student who has written the correct answer receives a point.

WHO: Grade 3–Grade 5
WHEN: During the lesson
FOCAL POINT(S): Measurement

- Place students in two heterogeneous teams. Have each team compete against the other as members take turns providing the appropriate unit of measurement while you read situations similar to the following:

  o How much a person weighs
  o How long a baseball game lasts
  o How far it is from home to school
  o How much Coke a student drinks in one day
  o How fast a person drives his car
  o How much water there is in the swimming pool
  o How much milk there is in the refrigerator
  o The size of a student's computer
  o The width of the school
  o How much gold a wedding band contains

Be sure to add items from the complied list which students bring in as the homework activity outlined in Chapter 3: Field Trips. The team with the most correct answers at the end of the game wins.

WHO: Grade 6–Grade 8
WHEN: During the lesson
FOCAL POINT(S):   Geometry

- Have students play the Geometry Loop game by writing the statements and questions (similar to the following) on index cards and passing them out randomly to students in the class. Students then use the cards to answer one another's questions.

  o **I have a right triangle.** Who has a triangle with all sides congruent?
  o **I have an equilateral triangle.** Who has the number of degrees in each of its angles?
  o **I have 60 degrees.** Who has the segment of a triangle from a vertex to the midpoint of the opposite side?
  o **I have median.** Who has a triangle with each angle less than 90 degrees?
  o **I have an acute triangle.** Who has a triangle with at least two congruent sides?
  o **I have an isosceles triangle.** Who has an equation whose graph is a line?
  o **I have a linear equation.** Who has the name of the side opposite the right angle in a right triangle?
  o **I have the hypotenuse.** Who has an equation for the area of a circle?
  o **I have a = πr^2.** Who has an equation that states that two ratios are equal?
  o **I have proportion.** Who has a quadrilateral with four congruent sides?

Students can write additional questions and answers that will form the basis of the remaining cards for playing this game. You should have as many cards as there are students in class (Bulla, 1996).

WHO: Grade 6–Grade 8
WHEN: During the lesson
FOCAL POINT(S):   Geometry

• Have students stand and play Simon Says by having students follow directions similar to the following:

> *Simon says: Do a 90-degree clockwise turn.*

> *Simon says: Do a 360-degree counterclockwise turn.*

> *Now do a 180 degree clockwise turn.*

Just as the rules of the game dictate, students should follow directions only when the words *Simon Says* precede the direction.

(Darla R. Quinn, Grade 5, Janet Johnstone Elementary School, Calgary, Alberta, Canada)

WHO: Grade 6–Grade 8
WHEN: During the lesson
FOCAL POINT(S):   Number and Operations

• Have eight students, representing eight various countries around the world (such as India, China, United States, Canada, Japan, and so forth) sit at different workstations. Divide the rest of the class into small groups. Have the groups rotate around the room (at the sound of a bell) to different countries where they must solve a particular exchange rate question at the station. The representative of each country can check the work for that group. If they get the question right, each member of the group receives two currency tokens for each correct answer. The group with the most currency tokens in the end is the winner!

(Sonia Duggal, Woodstock Collegiate Institute, Thames Valley District School Board, London, Onatrio, Canada)

WHO: Kindergarten–Grade 8
WHEN: After the lesson
FOCAL POINT(S):   All

• Create 15 math problems and their answers regarding a concept just taught. Write the problems on 15 separate index cards and place them in a bag. Play Bingo in the customary way by having students draw a 3×3 bingo card (a matrix with three boxes across and three boxes down, or nine boxes in all) on their paper. The nine squares should be large enough to write an answer to each math problem inside. Have students write the answers to the nine problems, randomly on their cards. (There is no "Free Space.") Have one student randomly draw a card from the bag and read the problem while all students place an "X" or a removable marker over the answer that accompanies each problem. The first student to get Xs on all four corners and shouts *Bingo* wins the game. Vary the game by declaring the winner as the student who is the first to cover three spaces in a row either up and down, across, or diagonally or the first to cover all nine spaces on the entire card.

WHO: Grade 3–Grade 8
WHEN: During the lesson
FOCAL POINT(S):   All

• To review math content previously taught, have students play the following game. Have them construct multiple-choice questions by selecting key points in a designated chapter. Collect and compile the questions. Draw names randomly for a student to enter the "hot seat." Ask students questions similar to those on the show, *Who Wants to be a Millionaire?* The student in the hot seat has three lifelines: 50/50 (eliminate two wrong answers out of four), phone a friend (ask a friend in class), and poll the audience (use raised hands in the classroom). The student who makes it to a predetermined level of questioning receives a small prize.

(Sonia Duggal, Woodstock Collegiate Institute, Thames Valley District School Board, London, Onatrio, Canada)

WHO: Grade 6–Grade 8
WHEN: During the lesson
FOCAL POINT(S):   Number and Operations

• Give students opportunities to practice their addition while solving problems and discovering patterns. Have them create a Magic Square by placing the numerals 1, 2, 3, 4, 5, 6, 7, 8, and 9 in three columns of three. Have them work individually or with a partner to arrange the numbers so that when you add them vertically, horizontally, and diagonally, the answer will continue to be 15. Two examples follow:

Example 1:   8   3   4
1   5   9
6   7   2

Example 2:   6   1   8
7   5   3
2   9   4

WHO: Grade 3–Grade 8
WHEN: During the lesson
FOCAL POINT(S):   All

• Encourage students to use appropriate math vocabulary by playing Pictionary. Divide the class into two heterogeneous teams. Students from each team take turns coming to the front of the room, pulling a vocabulary math word from a box, and drawing a picture on the board that will get their team members to say the word before time is called. No words may be spoken. If the team succeeds in guessing the word within a specific time limit (such as 30 seconds), the team gets one point. The team with the most points after all words have been used is the winner. The following vocabulary math words can be used: *data, survey, axis, frequency, bar graph, line graph, pictograph, range, median, mode, mean, interest, principal, spreadsheet.*

WHO: Grade 6–Grade 8
WHEN: After the lesson
FOCAL POINT(S): Probability

• Have students play the television game show *Deal or No Deal* in your classroom. Post 26 numbered cases on the board or wall, with the numbers 1–26 on one side of each case and an amount of money (from $.01 to $1 million) on the back. Post-it Notes work well for this purpose. Have class members select a "money case" from the 26 cases posted on the board. This becomes the "class case." Then have students take turns selecting a case from the board. You become the banker and offer students money to stop opening more cases and take the unknown amount in their chosen case. Ask each student selecting a case whether he or she wants to make a deal. As more cases are opened and eliminated, have students calculate the probability that their case has the $1 million. The game proceeds following the rules for playing the television show.

(Sonia Duggal, Woodstock Collegiate Institute, Thames Valley District School Board, London, Onatrio, Canada)

WHO: Grade 6–Grade 8
WHEN: During the lesson
FOCAL POINT(S): Algebra

• Have students solve algebra problems by playing indoor baseball. Prior to starting the game, the students or teacher should write math facts on laminated circles with 1st base, 2nd base, 3rd base, or HR (home run) written on the opposite side. The notation of base numbers will determine the complexity of the problem with a HR (home run) being the most difficult equation. Convert the class into a baseball diamond by labeling areas on the wall for students to move to as they respond to the equations.

Place the cards/circles in a central location with the equation side down. Divide the class into heterogeneous teams and use a coin toss to determine which team will start the game. The group who wins the toss is at bat first.

Option 1: Separate the cards into stacks according to the base number listed on the back of each card. The first team member must select a card from one of the stacks, read the equation aloud, and state the correct response to the problem. If the answer is correct, the student may advance to the appropriate base—as noted on the card. Record the scores as students reach home plate. The team members will continue to respond to equations until all team members have a chance to answer a problem or someone from the team responds incorrectly. In either case, the next team is up to bat.

Option 2: The other option is to place the cards in a mixed stack and follow the same procedure of allowing a team to respond until each player has had a chance to answer the equation or until the first person gets a problem incorrect.

This game can be adapted to any math skill simply by changing the problems written on the bases.

(Tomiko T. Smalls, Grade 2 Teacher, Mossy Creek Elementary School, North Augusta, SC)

WHO: Grade 6–Grade 8
WHEN: After the lesson
FOCAL POINT(S):   Number and Operations

• Old Maid Math: Make pairs of cards with an equation on one card and the graph of the equation on a matching card. Make 11 pairs of cards (22 cards total) and one "wild card." Play the game according to the following directions. Shuffle the cards and deal them out to all of the players. Depending on the number of players, some players might have more cards than others. Have players take turns picking one card at a time from another player. When a player has a match, the match is placed face up in front of the player. When all matches are made, the game is over. Every match is worth one point. The player with the "wild card" at the end of the game loses two points.

Alternate form of game: Use any type of pairs, such as equations and solutions, vocabulary words and their definitions, or fractions, decimals, and percent equivalents, to make the cards.

(Connie Matchell, Geometry and Algebra II Teacher, Siloam Springs Public Schools, Siloam Springs, AR)

WHO: Grade 6–Grade 8
WHEN: During the lesson
FOCAL POINT(S):   Algebra

• Divide the class into groups. Give each group an erasable whiteboard (like a mini-chalkboard) on which they can scribble and erase their answers. Put a problem or question up on the board and give the class three minutes to solve the problem. The first group with the correct answer (with all the necessary steps) wins the points for that round. A sample problem follows:

Collect like terms: $5x - 2a + 3x - 15b - 28x + 3a - 13b$

Solution: $-20x + a - 28b$

REFLECTION AND APPLICATION

> How will I incorporate *games* into
> mathematics instruction with my students?

Concept _____

Activity _____

_____

_____

Concept _____

Activity _____

_____

_____

Concept _____

Activity _____

_____

_____

Concept _____

Activity _____

_____

_____

Concept _____

Activity _____

_____

_____

Graphic Organizers, Semantic Maps, and Word Webs

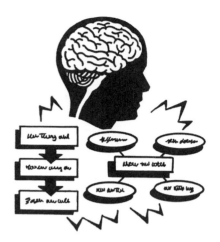

WHAT: DEFINING THE STRATEGY

During the 1990s, I was first introduced to the concept of graphic organizers, and my teaching has never been the same. As a former reading specialist, I also used thinking maps, word webs, and semantic or concept maps to teach a myriad of comprehension skills.

When instructing in each of the nine classes I teach today, I draw visuals called graphic organizers when I need to relate specific concepts that may be complex or difficult to comprehend. I have participants draw the organizer along with me as I talk about it. Then, to show the power of this strategy, I ask participants to turn their paper face down and ask them questions about the content just taught. Invariably, they get all of the answers correct when giving choral responses. Then we discuss the fact that, without the actual drawing of the visual on the part of the participants, comprehension would not be nearly as good. We also discuss the fact that if I had given them the organizer already printed on a handout, comprehension would also be compromised.

Math teachers can also use this time-tested strategy to teach to both left and right hemispheres of the brain. The left hemisphere, which tends to be more verbal, can see the words of the organizer while the right hemisphere, which tends to be more visual, can see the connections in the concepts represented.

WHY: THEORETICAL FRAMEWORK

Graphic organizers not only gain the attention of students but can also improve comprehension, meaning, and retention (Sousa, 2007).

Graphic organizers enable English learners to organize words and ideas in a way that helps them see patterns and relationships in mathematics (Coggins et al., 2007).

Flow charts, continuums, matrices, Venn diagrams (a type of graphic organizer), concept maps, and problem-solution charts are all types of graphic representations that can be used by mathematics teachers because they can be quickly understood and can provide structure for synthesizing new information (Posamentier & Jaye, 2005).

Concept maps, a type of graphic organizer, integrate both visual and verbal activities and enhance comprehension of concrete, abstract, verbal, and nonverbal concepts (Sousa, 2006).

Graphic organizers are powerful tools for instruction because they enable students to organize data into segments or "chunks" that they can comprehend and manage (Gregory & Parry, 2006, p. 198).

When graphic organizers are used to change words into images, both left- and right-brain learners can use those images to see the big picture (Gregory & Parry, 2006).

Graphic organizers are effective tools for supporting thinking and learning in four major ways: (1) abstract information is represented in a concrete format; (2) relationships between facts and concepts are depicted; (3) new information is connected to previous knowledge; and (4) thoughts are organized for writing and for problem solving (Ronis, 2006).

When teachers use graphic organizers, learning is more easily understood and remembered since knowledge is arranged into "holistic, conceptual frameworks" (McTighe, 1990, p. 33).

HOW: INSTRUCTIONAL ACTIVITIES

WHO: Grade 1–Grade 3
WHEN: During the lesson
FOCAL POINT(S): Number and Operations

- Draw a graphic organizer on the board similar to the one below. In the middle of the web, write a number. In the six circles write as many ways as possible in which that number can be represented. For example, if the number is 10, the six circles could include (6 + 4), (5 + 5), (8 + 2), (13 – 3), (20 – 10), and (10 – 0).

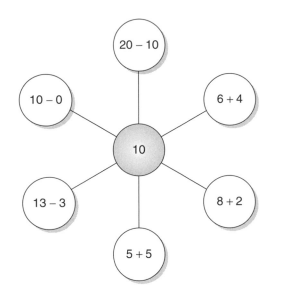

WHO: Prekindergarten–Grade 2
WHEN: During the lesson
FOCAL POINT(S): Number and Operations

• Draw graphic organizers on the board to provide students with pictorial representations of how the ideas are related in a word problem. For example, the following word problem could be pictured below:

Christopher has two dogs. Jessica has seven cats. How many animals do Christopher and Jessica have together?

WHO: Grade 3–Grade 8
WHEN: During the lesson
FOCAL POINT(S): All

• Have students use the following Word Problem Map (Bender, 2005, p. 80) when scaffolding their comprehension of simple word problems.

A Word Problem Map

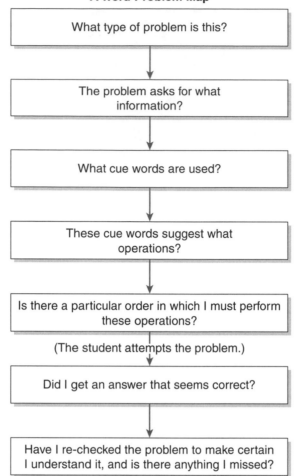

WHO: Grade 7
WHEN: During the lesson
FOCAL POINT(S):   Probability

• Pose a problem similar to the following: Michael has two hamsters (H1 and H2) and three cages (C1, C2, and C3). He can choose any hamster to place in any cage. What is the probability that he will choose to place H1 into C2? Have students draw and complete the following graphic organizer to help them see the data more clearly and answer the question.

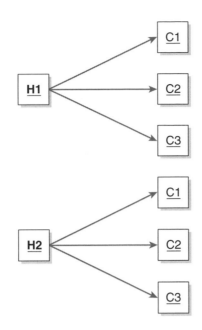

WHO: Grade 3–Grade 8
WHEN: During the lesson
FOCAL POINT(S): All

• Have students select the appropriate graphic organizer (chart, table, or graph) to depict data gathered during a project.

WHO: Grade 3–Grade 8
WHEN: During the lesson
FOCAL POINT(S): Algebra

• Use a thinking, mind, or semantic map to show students the steps in solving algebraic equations. Have them copy the map in their notes for future reference. Provide students with other opportunities to solve problems using the same organizer. For those who are having difficulty, fill in part of the organizer with the correct steps in solving the equation, while leaving other steps for the student to complete. Students may also work in pairs to facilitate understanding.

WHO: Grade 3–Grade 8
WHEN: During the lesson
FOCAL POINT(S): Geometry

• Have students use a graphic organizer or Venn diagram to compare and contrast two geometric shapes.

WHO: Grade 3–Grade 5
WHEN: During the lesson
FOCAL POINT(S): Number and Operations, Measurement

• Design a graphic organizer that shows the relationship of standard units of measurement to the units of measurement in the metric system and post it on the wall as a visual.

WHO: Grade 6–Grade 8
WHEN: During the lesson
FOCAL POINT(S):   Algebra

- To assist students in simplifying multistep algebraic equations, use a graphic organizer similar to the one below:

Algebraic equation:
Use the distributive property:
Combine like terms:
Undo addition and subtraction:
Undo multiplication and division:
Answer:

(Tate, 2008).

Graphic organizers similar to the one above can be used to delineate the steps in solving any multistep problem.

WHO: Grade 6–Grade 8
WHEN: During the lesson
FOCAL POINT(S):   Algebra

- Design a graphic organizer, called a word web, regarding any math vocabulary concept. A sample of the word web is depicted in the first activity of this chapter. Rather than a number, the name of the concept is written in the middle circle and ideas related to the concept are written in surrounding circles. For example, if the word *triangle* is written in the middle of the circle, the following words: *3 sides, 3 angles, equilateral, scalene,* and *isosceles* could be written in the surrounding circles. The word web can be posted on the wall as a visual while students draw it in their notes for later reference.

WHO: Grade 3–Grade 6
WHEN: During the lesson
FOCAL POINT(S):   Algebra

- Select word problems that engage students in making a table or chart to organize information in order to craft a solution to the problem.

REFLECTION AND APPLICATION

> **How will I incorporate *graphic organizers*, *semantic maps*, and *word webs* into mathematics instruction with my students?**

Concept _____

Activity _____

_____

_____

Concept _____

Activity _____

_____

_____

Concept _____

Activity _____

_____

_____

Concept _____

Activity _____

_____

_____

Concept _____

Activity _____

_____

_____

STRATEGY **6**

Humor

WHAT: DEFINING THE STRATEGY

Why should you never date a tennis player? (To them *love* means nothing.)

What do you get if you cross a turtle and a porcupine? (You get a *slow poke*.)

How much do pirates' earrings cost? They cost a *buccaneer*. (*a buck an ear*)

Riddles, jokes, and funny stories have the ability to minimize stress and maximize achievement. When students are laughing *with* each other, they don't have time to laugh *at* each other. The National Council of Teachers of Mathematics knew this almost 40 years ago when in 1970 they published a book called *Mathematics and Humor*. Every riddle, joke, and cartoon in this book teaches or reinforces a math concept. Watch students' love of math and retention of complex concepts increase when they are taking pleasure from their learning.

Every class has a *class clown*; sometimes it is even the teacher! Use this personality to your advantage. Share riddles, jokes, or cartoons when the class needs some downtime or a "pick-me-up" for the brain. Have students think at higher levels as they design riddles, jokes, and cartoons for their classmates to enjoy. After all, what do you call a fish that operates on the brain? (*a neurosturgeon*)

WHY: THEORETICAL FRAMEWORK

Having students set and achieve personal goals, such as learning their multiplication tables, helps them build on their successes and savor memories of positive feelings (Willis, 2007).

Since laughter increases the endorphin level in the body, lowers the heart rate and the stress hormone cortisol, students in a laughing classroom are more likely to take risks and tend to be more creative and collaborative (Gregory & Parry, 2006).

Play is essential in helping children practice those social skills essential for everyday life (McCormick Tribune Foundation, 2004).

When students are experiencing minimal stress, levels of cognition are increased and information is allowed to flow more freely through the amygdala, the seat of emotion (Willis, 2007).

Older adolescents are more apt to understand the subtleties of humor, satire, or irony because their language skills are more highly developed than those of younger students (Feinstein, 2004).

The use of humor is one of 12 intelligent behaviors, labeled as *habits of mind*. These *habits* are based on the premise that all students can be taught a set of skills that enable them to behave in intelligent ways (Costa, 1991).

Cows may be contented . . . but only men and women, girls and boys laugh. The human mind alone can see the incongruous amid the congruities of life (National Council of Teachers of Mathematics, 1970).

HOW: INSTRUCTIONAL ACTIVITIES

WHO: Prekindergarten–Grade 8
WHEN: After the lesson
FOCAL POINT(S):   All

- Have students create riddles regarding math concepts for their classmates to solve. For example, two sample riddles might be:

  1. *Question:* There are only eight of these in a container. What are they? (Students should look around the room until they find the eight pencils in a can on the table.)

  2. I am less than 100.

     Three and 27 are two of my factors.

     Two is not one of my factors.

     I am a square number.

     Which number am I?

     Answer: 81

  3. *Question:* Why is simplifying a fraction like powdering your nose?

     Answer: It improves the appearance without changing the value.

(National Council of Teachers of Mathematics [1970])

WHO: Grade 3–Grade 8
WHEN: During the lesson
FOCAL POINT(S): All

- Have students design cartoons or funny stories that demonstrate their understanding of a mathematics concept such as the following:

WHO: Grade 3–Grade 5
WHEN: During the lesson
FOCAL POINT(S): Number and Operations

- Have students solve the following riddle: Have them pick any number and then add the next larger number to it. Then have them add 9 to the sum, whatever it is. Then they should divide the result by 2 and subtract the original number. The answer will be 5.

> For example, the number selected is 15.
>
> Adding the next number to it will give you 15 + 16 = 31
>
> Adding nine to 31 will give you 40.
>
> Dividing by 2 will give you 20.
>
> Subtracting the original number 15 will give you 5, the correct answer.

Students can then try other numbers to see if the process holds true. It will! Have them work with a partner to figure out why (Bulla, 1996).

WHO: Grade 6–Grade 8
WHEN: After the lesson
FOCAL POINT(S): Algebra, Measurement, Number and Operations

- Have students create riddles regarding a math concept previously taught. Some examples appear on the following page.

Riddle: What do you call a mermaid's undergarment?

Answer: An algae bra.

(National Council of Teachers of Mathematics [1970])

Riddle: Why did they put the mathematician in prison?

Answer: He tried to kilo meter.

(National Council of Teachers of Mathematics [1970])

Riddle: What did the number 0 say to the number 8?

Answer: Where did you get that belt?

(Reprinted with permission from *Mathematics and Humor,* edited by Aggie Azzolino, et. al. [1970] by the National Council of Teachers of Mathematics. All rights reserved.)

WHO: Grade 6–Grade 8
WHEN: After the lesson
FOCAL POINT(S):   All

• Have students create original jokes regarding a math concept previously taught. The creation of jokes not only reinforces students' conceptual understanding but also encourages students to use their higher-level thinking skills.

WHO: Grade 3–Grade 8
WHEN: Before the lesson
FOCAL POINT(S):   All

• Have students take turns bringing in jokes. Place appropriate jokes in a box. At the beginning of each period or each day, select one joke to tell to the class. One middle school teacher related to me that when she started this procedure, her percentage of students coming to class on time improved considerably. Her students anticipated the positive start to class.

WHO: Prekindergarten–Grade 8
WHEN: After the lesson
FOCAL POINT(S):   All

• Play games with students to review content prior to a test. Consult Chapter 4: *Games* for numerous examples for involving students in a fun math class with lots of humor.

WHO: Grade 3–Grade 8
WHEN: After the lesson
FOCAL POINT(S):   All

• Locate newspaper editorials or cartoons which emphasize math concepts already taught. Display them in class and have students use their higher-level thinking skills to explain the concept displayed in the cartoon. For older students, you may want to display the cartoon omitting the caption and have students work individually or in groups to create their own captions. You would be surprised how your students' original captions may be superior to the ones provided in the cartoons.

WHO: Grade 3–Grade 8
WHEN: After the lesson
FOCAL POINT(S):   All

- Have students design original cartoons, comic books, or superheroes to illustrate a key math concept taught. For example, students could design a comic book where the main character is Polygon, a superhero with all the strengths and powers of a polygon (Tate, 2006).

WHO: Grade 6–Grade 8
WHEN: After the lesson
FOCAL POINT(S):   All

- Guide students into developing an episode of a "math sitcom." Some students can be the directors, some can be the writers, and some can be the actors. All of the students can contribute to the creative production. This will give them an opportunity to try their math jokes "live" as they perform the sitcom for an audience.

REFLECTION AND APPLICATION

How will I incorporate *humor* into
mathematics instruction with my students?

Concept _____

Activity _____

_____

_____

Concept _____

Activity _____

_____

_____

Concept _____

Activity _____

_____

_____

Concept _____

Activity _____

_____

_____

Concept _____

Activity _____

_____

_____

Manipulatives, Experiments, Labs, and Models

WHAT: DEFINING THE STRATEGY

I observed in one classroom where students were learning to identify and label various types of angles—acute, right, and obtuse. In the front of the classroom, the teacher placed a box full of a variety of objects that could be used as manipulatives: toothpicks, pipe cleaners, and straws, to name a few. Students worked in cooperative groups to make designated angles using only one type of object. Not only did the class have fun, they were actively engaged in using their hands to complement what their brains were learning. It worked! When tested on the same objective, every student earned a grade of *B* or better.

The strategy of manipulatives and model building may be the one strategy most closely associated with meaningful math instruction. There is such a strong connection between the hands and the brain that you will see many students counting on their fingers before they no longer need them. You will see classrooms where Unifix cubes and Geoboards are the order of the day.

Students with what Howard Gardner (1983) calls *visual-spatial intelligence* may have difficulty when given a math worksheet but may excel when using their hands to solve a rubrics cube or construct a model. Students with this gift also tend to do well in geometry. Geometry has a major place in the mathematics curriculum because it is the best tool we have for describing, analyzing, and understanding the physical world in which we live. Geometry also provides support for the study of other areas of the math curriculum such as number sense, measurement, and algebra.

WHY: THEORETICAL FRAMEWORK

When students are working with concrete shapes, they are developing the foundation for spatial sense (Wall & Posamentier, 2006).

Students' understanding of mathematical ideas is broadened when concrete representations are used (Coggins et al., 2007).

When learning is active and hands-on, the formation of neural connections is facilitated and information is much more readily remembered than information learned from an abstract viewpoint, where the teacher is doing the work while the students watch (Gregory & Parry, 2006).

Since concrete materials assist English language learners in focusing on new concepts and vocabulary at the same time, they are a crucial part of the instruction in fluency with mathematics (Coggins et al., 2007).

In prekindergarten though Grade 2, students can best learn about geometry through using objects that they can see, hold, and manipulate (National Council of Teachers of Mathematics, 2000).

The use of the hands and brain activity are so complicated and interconnected that no one theory explains it (Jensen, 2001).

Students in the early grades should be allowed to use manipulatives for as long as the students feel they are needed (Checkley, 1999).

The best teaching techniques for fostering intelligence unite rather than separate the body and the mind (Wilson, 1999).

HOW: INSTRUCTIONAL ACTIVITIES

WHO: Prekindergarten–Grade 1
WHEN: During the lesson
FOCAL POINT(S):   Number and Operations

• Have students correlate place value with counting by 10 as they use either Unifix cubes, Popsicle sticks, base 10 blocks, or other similar items. Accompany these manipulatives with the following number table.

Number Table

1	2	3	4	5	6	7	8	9	10
11	12	13	14	15	16	17	18	19	20
21	22	23	24	25	26	27	28	29	30
31	32	33	34	35	36	37	38	39	40
41	42	43	44	45	46	47	48	49	50
51	52	53	54	55	56	57	58	59	60
61	62	63	64	65	66	67	68	69	70
71	72	73	74	75	76	77	78	79	80
81	82	83	84	85	86	87	88	89	90
91	92	93	94	95	96	97	98	99	100

Using the manipulatives and the numbers table, have students answer questions such as the following:

o Show me 10. Show me 10 more.
o Show me 40. Take 10 away. How many are left?

WHO: Prekindergarten–Grade 3
WHEN: During the lesson
FOCAL POINT(S): Measurement

• Have students use standard and non-standard units (i.e. paper clips) to measure objects in and around their school environment, such as their desks. Have them use mathematical terminology such as taller than, wider than, shorter than, higher than, deeper than, when comparing two or more objects measured.

WHO: Grade 1–Grade 3
WHEN: During the lesson
FOCAL POINT(S):   Number and Operations

- Give each student a miniature clock with moveable hands. As you state a time, have students show that time on their miniature clocks. Start with times that are on the hour and half hour and move into more complicated times.

WHO: Grade 1–Grade 3
WHEN: During the lesson
FOCAL POINT(S):   Geometry

- Have students work in groups of four to six. Give each group an envelope containing cardboard copies of the multiple shapes that have been studied. Have them work together to sort the shapes by category—circle, square, triangle, rectangle, and so forth.

WHO: Prekindergarten–Grade 1
WHEN: During the lesson
FOCAL POINT(S):   Geometry

- Give each student a plastic bag with assorted stickers. Have them sort the stickers into groups based on a specified criteria, such as same size, same color, and same design. Then have the stickers arranged by size such as smallest to largest (National Council of Teachers of Mathematics, 2000).

WHO: Grade 3
WHEN: During the lesson
FOCAL POINT(S):   Number and Operations

- Students can use their fingers as a human calculator for the nine times table. Have students hold their hands up in front of them (palm side facing away from the body) and spread their fingers. Starting with the little finger on the left hand, have students assign a number to each finger. When multiplying using the nine times table, have them turn down the finger which represents the number to be multiplied times nine. For example, if the student is multiplying nine times four, then the fourth finger on the left hand is turned down. All fingers to the left of the turned down finger is the first digit in the answer, and all fingers to the right of the turned down finger is the second digit in the answer. Therefore, the answer to $9 \times 4 = 36$.

WHO: Grade 3–Grade 5
WHEN: During the lesson
FOCAL POINT(S):   Number and Operations

- Give each student three strips of paper of the same length. Have them learn about equivalent fractions by folding their sets of papers into three different fractional parts—in halves, in fourths, in eights. Help them to conclude that all three strips are equal in length although they are folded differently.

WHO: Grade 3–Grade 5
WHEN: During the lesson
FOCAL POINT(S):   Number and Operations

• Provide students with manipulatives to assist them in understanding concepts in math problems. For example, have each student bring a Kit Kat bar to class. This bar consists of four naturally divided segments and is very useful for teaching fractional parts. Give each student a plastic knife and have them cut off one segment (1/4) of the bar. Have them further divide that one segment into halves; the next segment into thirds; and the next segment into fourths. The last segment could be divided into sixths. Then have them give away 3/6 of the last segment. Continue having students divide and give away segments of the candy bar until they have some understanding of fractional parts of a whole.

(Jennifer Clowers, Grade 2 Teacher, Bob Mathis Elementary School, Decatur, GA)

WHO: Grade 3–Grade 5
WHEN: During the lesson
FOCAL POINT(S):   Geometry

• Have students make geometric shapes in the air with their fingers or model angles (acute, obtuse, right) with their arms.

WHO: Grade 7
WHEN: During the lesson
FOCAL POINT(S):   Probability

• Have students begin to understand probability by having them work in pairs to roll two number cubes. While one student rolls, the other one writes down the two numbers generated. They then add the two numbers together. Have them predict which numbers will be rolled more often. They should begin to realize that some sums (6, 7, 8) will occur more often than others (2, 12) (National Council of Teachers of Mathematics, 2000).

WHO: Grade 7
WHEN: During the lesson
FOCAL POINT(S):   Probability

• Replicate the activity listed above using spinners rather than number cubes. Have students make predictions regarding which sums will appear most often.

WHO: Grade 3–Grade 5
WHEN: During the lesson
FOCAL POINT(S):   Geometry

• Have students work in groups of four to six. Provide each group with a set of blocks that snap together. Have each group work with one another to build a three-dimensional figure that is several stories high.

When the figures are complete, place them at an angle on a table or desk where the class can only see a front and slight side view. Have students answer the following questions regarding each figure:

- How many blocks in the figure are hidden from view?
- How many total blocks are in the figure?
- How many blocks would be hidden from view if we made the structure three blocks taller? Four blocks wider?

WHO: Grade 4–Grade 5
WHEN: During the lesson
FOCAL POINT(S):   Measurement

- Have students cut a piece of string equal to their height. Each student compares the length of the string with their reach. Ask students whether they are rectangles or squares. Have them use the string to figure out other ratios such as, "Is your foot equal to putting the string once around your wrist? Is once around the neck equal to twice around your wrist?" (Burns, 1975, p. 37).

WHO: Grade 7
WHEN: During the lesson
FOCAL POINT(S):   Probability

- Have students work in groups. Give each group the following materials:

- A small paper bag with 100 small beads of three different colors (put the same number of colored beads in each bag, such as 30 red, 50 white, and 20 blue)
- A group tally table
- A class summary table
- A four-function calculator
- A sharp pencil

Without looking inside the bag, one student in each group should reach into the bag and pull out exactly 10 beads. (All students should note the colors of the 10 beads selected.) Another student in each pair records the results of each trial on the group tally table. Have students repeat the process 10 times. They should return the beads to the bag after each selection. Students should calculate the sum of each color and record on the group tally table. Have students make a prediction. Given the information they have on their group tally tables, have them predict how many of each color of beads are in the bag. Then have students dump the beads out to be hand counted and see how close their predictions were. A copy of the group tally table and the class summary table follows.

Group #	Red	White	Blue
1			
2			
3			
4			
5			
6			
7			
8			
9			
10			
Totals			
Probability %			

Group #	Red	White	Blue
1	53	28	19
2	48	31	21
3	54	24	22
4	51	33	16
5	49	31	20
6	47	32	21
7	50	27	23
8	53	30	17
9	46	35	19
10	52	25	23
Totals	503	296	201
Probability %	= 50.3 or **Approx. 50%**	= 29.6 or **Approx. 30%**	= 20.1 or **Approx. 20%**

WHO: Grades 3–5
WHEN: During the lesson
FOCAL POINT(S):   Geometry

• Have students use attribute blocks, tangrams, Geoboards, or pattern blocks to help them perceive the difference in patterns of geometric figures. For example, they can use a Geoboard to form triangles, squares, parallelograms, and so forth.

WHO: Grade 7
WHEN: During the lesson
FOCAL POINT(S):   Probability

• Have students work with a partner to make predictions as to the number of times a specific coin will show up either heads or tails when flipped 10 times. Then have them take turns flipping the coin and recording the data. Have them work with five sets of 10 flips each and determine whether there is a pattern in the number of heads and tails obtained. Have them then compare their answers to the data from another pair of students. You may want to compile class data and have students discover that the more times the coins are flipped, the closer the frequency of heads and tails is to 50/50.

WHO: Grade 8
WHEN: During the lesson
FOCAL POINT(S):   Measurement, Geometry

• Students build scale models of the house that they hope to purchase as an adult. They must decide which rooms to include in their house and the dimensions of each room. They must then construct a scale model. For example, one square centimeter equals one square meter.

(Darla R. Quinn, Grade 5, Johnstone Elementary School, Calgary, Alberta, CA)

WHO: Grade 3–Grade 8
WHEN: During the lesson
FOCAL POINT(S):   Algebra, Measurement

• Support interdisciplinary instruction by talking with the science teachers of your students and find out if students will be given opportunities to analyze data in upcoming class experiments. Find out which processes or formulas will be necessary for analyzing the data; then teach those in math class. The more students can see the connections between content areas, the more sense the content makes to their brains.

WHO: Grade 6–Grade 8
WHEN: During the lesson
FOCAL POINT(S): Algebra

• Place the steps for solving six algebraic equations on six different colors of construction paper and place them in six different envelopes. Have students work in cooperative groups to demonstrate their understanding of solving these equations by placing the paper steps in order from the first to the final answer. Have each group compare their solution to the real steps in the correct order. An example follows:

1. *Steps in Solving an Algebraic Equation*
   o $6x - 10 = 4x + 26$
   o $6x - 4x - 10 = 4x + 26 - 4x$
   o $2x - 10 = 26$
   o $2x - 10 + 10 = 26 + 10$
   o $2x = 36$
   o $2x/2 = 36/2$
   o $x = 18$

Place these steps on separate strips of colored paper and put them in an envelope for one group to solve. Create five other envelopes. Have groups rotate the envelopes so that each group gets to experience six problems.

(Sonia Duggal, Woodstock Collegiate Institute,Thames Valley District School Board, London, Ontario, Canada)

WHO: Grade 6–Grade 8
WHEN: During the lesson
FOCAL POINT(S): Algebra

• Have students use area models to show the commutative property of multiplication. Have them place tiles like the ones below on their desks to demonstrate that 2×6 is the same as 6×2.

2×6

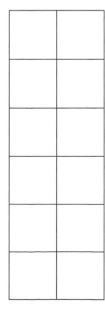

6×2

WHO: Grade 6–Grade 8
WHEN: During the lesson
FOCAL POINT(S): Algebra

- Have students use area models to demonstrate the distributive property of multiplication. Have them place tiles like the ones below on their desks to demonstrate that 7×16 is the same as 7×10 and 7×6.

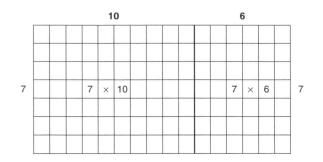

WHO: Grade 6–Grade 8
WHEN: During the lesson
FOCAL POINT(S): Number and Operations

- Guide students in using the principles of the Sieve of Eratosthenes to find all of the prime numbers between 1 and 100. Circle the first and smallest prime number (2). Cross out (1) because it is not prime by default of the definition of prime numbers. Then cross out every number divisible by (2). Circle the next prime number (3). Then cross out every number divisible by (3). Continue until all of the possible crossing out is done. The remaining numbers represent the prime numbers between 1 and 100.

Number Table

1	2	3	4	5	6	7	8	9	10
11	12	13	14	15	16	17	18	19	20
21	22	23	24	25	26	27	28	29	30
31	32	33	34	35	36	37	38	39	40
41	42	43	44	45	46	47	48	49	50
51	52	53	54	55	56	57	58	59	60
61	62	63	64	65	66	67	68	69	70
71	72	73	74	75	76	77	78	79	80
81	82	83	84	85	86	87	88	89	90
91	92	93	94	95	96	97	98	99	100

REFLECTION AND APPLICATION

> ## How will I incorporate *manipulatives, experiments, labs,* and *models* into mathematics instruction with my students?

Concept _____

Activity _____

_____

_____

Concept _____

Activity _____

_____

_____

Concept _____

Activity _____

_____

_____

Concept _____

Activity _____

_____

_____

Concept _____

Activity _____

_____

_____

Metaphors, Analogies, and Similes

WHAT: DEFINING THE STRATEGY

Complete the following similes:

Like sands through the hour glass, _____ .

Like a good neighbor, _____ .

Life is like a box of chocolates, _____ .

If you have ever watched the soap opera *Days of Our Lives*, purchased insurance from State Farm, or seen the movie *Forrest Gump*, then you already know the answers to the aforementioned similes. If not, then ask someone who has.

When students are comparing two dissimilar concepts with the use of the words *like* or *as*, they are using similes. Comparisons without the use of *like* or *as* are considered metaphors. Analogies show resemblances between things that are otherwise not alike. *Math is to Pythagorus as Science is to Hippocrates* is an example of an analogy. Metaphors, similes, and analogies are instructional strategies that people use constantly in real life to help them comprehend complex concepts. Mathematics is no exception. Anytime I can take a new math concept that students do not understand and connect it to a concept that they do understand, it helps to clarify the new concept. This strategy is one of nine that Marzano, Pickering, and Pollack (2001) recommend as an instructional strategy that really works. Try it and you will see that we are right.

WHY: THEORETICAL FRAMEWORK

When students connect what they are learning in mathematics with other content areas, math is viewed as more useful and interesting than when math is taught as a separate subject (Posamentier & Jaye, 2005).

Using analogies to clarify or explain ideas assists students in making pertinent connections and increasing their comprehension of content (Gregory & Parry, 2006).

Teachers who give students analogies when providing explanations have students who are capable of conceptualizing complex ideas (Posamentier & Jaye, 2005).

Metaphors allow one to examine a concept from a broader perspective, such as how it applies across the curriculum, to the real world of the student, or to life as a whole (Allen, 2001).

Making metaphorical connections stretches students' thinking and increases their understanding (Gregory & Chapman, 2002).

Having students classify, compare, contrast, and use analogies and metaphors increases their achievement since they can look for similarities and differences between ideas or things (Marzano et al., 2001).

The majority of concepts are understood only in relation to other concepts (Lakoff & Johnson, 1980).

The concept of *synectics* is the use of analogies to make the familiar strange and the strange familiar. An example would be, "Why is the brain like a chain?" (Because it has many links and connections) (Gordon, 1961).

HOW: INSTRUCTIONAL ACTIVITIES

WHO: Grade 1–Grade 3
WHEN: During the lesson
FOCAL POINT(S):   Number and Operations

• Show students that numbers are analogous to other numbers or representations. For example, the number 36 can be represented in every one of the following ways: 20 + 16, three groups of 12, six groups of six, 50–14, and so forth.

WHO: Grade 1–Grade 4
WHEN: During the lesson
FOCAL POINT(S):   Number and Operations

• Assist students in understanding that certain operations in math are analogous to other operations. Help them to see that addition and subtraction are simply inverse operations, as are multiplication and division. Show them that multiplication is simply a faster form (more efficient way) of addition. The inverse of the fraction a/b is b/a as long as a is not (0). Because the brain is constantly searching for connections, consistently demonstrating these relationships to students should help them to better understand and apply these concepts.

WHO: Grade 1–Grade 3
WHEN: During the lesson
FOCAL POINT(S): Number and Operations

• When determining when to use the greater than (>) and the less than (<) signs, a simile is helpful. Put a green sock on your right hand and pretend that your hand is a sock puppet of an alligator. Tell students that when the alligator is hungry, it opens its mouth and eats or devours everything in sight. Open the mouth of your alligator puppet with its mouth turned to the left. Tell students that this mouth represents the greater than (>) sign since the great alligator can eat anything smaller than it is. Show several examples on the board, such as the following: 6 > 3, 15 > 13, 302 > 300, 1,244 > 1,241, and so forth.

WHO: Grade 6–Grade 8
WHEN: During the lesson
FOCAL POINT(S): Number and Operations

• Assist students in seeing the analogous relationship between types of notes in music and fractions in math. For example, in 4/4 time, it takes two half notes, four quarter notes, and eight eighth notes to make a whole note. Similarly, it takes, 2/2, 4/4, and 8/8 to also make a whole. Have students write musical notation to symbolize fractional parts.

WHO: Grade 6–Grade 8
WHEN: During the lesson
FOCAL POINT(S): Algebra

• To help students comprehend positive and negative slope, show them that this concept is analogous to the graphs on the stock page in the newspaper depicting whether the stocks are going up or down. Have students cut out sample graphs and label them as positive or negative slopes.

(Susan Sylvain Gallo, Grades 9–12, Black River High School, Renton, WA)

WHO: Grade 3–Grade 5
WHEN: During the lesson
FOCAL POINT(S): Geometry

• Have students look for examples of things in the real world that are indicative of symmetry or reflection, congruence, and other concepts. For example, discuss why the human body would be an example of symmetry.

WHO: Grade 3–Grade 5
WHEN: During the lesson
FOCAL POINT(S):   Geometry

• Have students look for examples of two- and three-dimensional representations of shapes in the real world. These would be objects or buildings that would be analogous to the geometric shapes studied. For example, a slice of pizza would be representative of a triangle; a stop sign representative of an octagon; the Pentagon in Washington, DC representative of a pentagon; an ice hockey rink representative of a rectangle, and so forth.

WHO: Grade 3–Grade 5
WHEN: During the lesson
FOCAL POINT(S):   Algebra

• Assist students in understanding algebraic equations by showing them a visual of a scale with two sides, which are balanced. Explain to them that when solving algebraic equations, those equations must also be balanced on both sides like the scale. Explain that what is placed on one side of the scale must be equal to what is placed on the other side. The same is true for the algebraic equation. Work some examples on the board to reinforce this simile.

WHO: Kindergarten–Grade 8
WHEN: During the lesson
FOCAL POINT(S):   All

• Show students that mathematics is naturally connected to all other areas of the curriculum by integrating it whenever possible as you are teaching language arts, social studies, science, art, music, or any other subject. For example, compare the patterns and rhythms in music with the patterns and rhythms in math. When conducting a science experiment, use math to derive the answer. When reading a story, such as *The Titantic*, compute the length of time that elapsed from the moment the ship initially hit an iceberg until the time it sank.

WHO: Grade 6–Grade 8 (Advanced)
WHEN: During the lesson
FOCAL POINT(S):   Algebra

• When students are overwhelmed by the sight of a complex equation (such as a polynomial with negative exponents) and have trouble visualizing it in its simplest form, use a colorful metaphor to maintain their attention. For example, relate the equation to fixing a sandwich for lunch. Consider each horizontal line to be the meat or fixings in the sandwich and the variables to be the bread. The equation below would likely be the strangest sandwich ever seen. It would consist of bread/meat/bread/meat:

$$\frac{\dfrac{b^2}{a^5 z^8}}{\dfrac{1}{a^6 b^3 z^{12}}}$$

Ask students how these two sandwiches can be broken up. Tell them that in the mathematical world an axiom exists that allows the multiplication of the reciprocal in the denominator to help simplify the equation.

When students are having difficulty conceptualizing the reciprocal of an equation, use another analogy: Ask them to conceptualize a trapeze artist swinging to and fro on a monkey bar. If the trapeze artist swings back and forth, his or her momentum will carry them to the next platform. Relate the equation to the swinging of a trapeze artist. Have students visualize the extra meat in the sandwich as a platform or hinge spot. If the trapeze artist swings to and fro, his momentum will bring him up to the next platform. However, the rope will dangle below him.

$$\frac{b^2}{a^5 z^8} \times \frac{a^6 b^3 z^{12}}{1}$$

The final step to simplifying the equation is to combine like terms (addition of exponents). Thus, the equation equals:

$$\frac{ab^5 z^4}{1} \text{ or } ab^5 z^4$$

(Chris Ginakos, Foundations for the Future Charter Academy, Dr. Norman Bethune Campus, Calgary, Alberta, Canada)

REFLECTION AND APPLICATION

How will I incorporate *metaphors*, *analogies*, and *similes* into mathematics instruction with my students?

Concept _____

Activity _____

_____

_____

Concept _____

Activity _____

_____

_____

Concept _____

Activity _____

_____

_____

Concept _____

Activity _____

_____

_____

Concept _____

Activity _____

_____

_____

STRATEGY 9

Mnemonic Devices

WHAT: DEFINING THE STRATEGY

Acrostics and acronyms are examples of mnemonic devices. Mnemonic devices are used in the real world all the time to help the public retain information. For example, most people are not going to remember *Acquired Immune Deficiency Syndrome*, so it is referred to as AIDS. *Sudden Infant Death Syndrome* is simply called *SIDS*, and *HOMES* stands for the Great Lakes of *Huron, Ontario, Michigan, Erie,* and *Superior.*

Please Excuse My Dear Aunt Sally (*PEMDAS*), has become the most well-known acrostic for helping students remember the order of operations when solving an algebraic equation. The acrostic reminds students to start by solving inside the *parentheses,* then simplifying the *exponents,* next *multiplying* or *dividing* (whichever comes first when looking from left to right), then finally *adding* or *subtracting* (whichever comes first from left to right). While mnemonics are not necessarily helpful for students as they comprehend complex mathematical concepts, they can assist them in retaining previously taught material. And they appear to work! After all, I learned *My Very Educated Mother Just Served Us Nine Pizzas* as an acrostic for the planets in order from the sun. Alas we have now lost Pluto, so I have changed it to *My Very Educated Mother Just Served Us Nachos!*

WHY: THEORETICAL FRAMEWORK

Process mnemonics, such as *PEMDAS (Please Excuse My Dear Aunt Sally),* are very effective for students having difficulty in math because they are attention-getting, motivational, and actively engage the brain in processes essential to learning and memory (Sousa, 2007).

Recall and retention can improve when teachers provide students with a mnemonic aid (Ronis, 2006).

> Mnemonics takes concrete associations and links them with abstract symbols making mathematics instruction cohesive and relevant (Bender, 2005).
>
> Japanese students with normal abilities, as well as those who are having difficulty in mathematics, can use process mnemonics to organize and summarize math problems (Manolo, Bunnell, & Stillman, 2000).
>
> While mnemonic techniques do not promote comprehension of mathematical concepts, they can assist in cueing and recalling connections to previously learned material (Ronis, 1999).
>
> Process mnemonics, schemes that help students remember, can be used to assist elementary students with operations and computation (Manolo, Bunnell, & Stillman, 2000).
>
> Content makes better sense to students when the structure of process mnemonics is used (Higbee, 1987).

HOW: INSTRUCTIONAL ACTIVITIES

WHO: Grade 3–Grade 8
WHEN: Before the lesson
FOCAL POINT(S): All

- To help eliminate the threat that occurs in the brain when some students lack confidence in their ability to be successful in your math class, teach them the acronym *MATH,* which actually stands for *Math Ain't That Hard!* Then use the 20 strategies as you teach and watch the acronym come true.

WHO: Grade 3–Grade 5
WHEN: During the lesson
FOCAL POINT(S):   Number and Operations

- Have students remember the acrostic *Does MacDonald's Sell Burgers?* to recall the steps in long division: **D**ivide, **M**ultiply, **S**ubtract, **B**ring down.

WHO: Grade 3–Grade 8
WHEN: Before the lesson
FOCAL POINT(S):   Number and Operations

- Teach students the *RIDD* strategy when learning to solve word problems. *RIDD* is an acronym for *Read* the problem from the beginning to the end; *Imagine* the problem by changing the new information into visual, auditory, and kinesthetic formats that have meaning to the brain; *Decide* what procedures to use to solve the problem; and *Do* the work needed to solve the problem (Jackson, 2002).

WHO: Grade 3–Grade 8
WHEN: During the lesson
FOCAL POINT(S):   Number and Operations

- To help students read the actual text of a word problem, use the *SQRQCQ* strategy. This strategy is a mnemonic device for the following:
    - **S**urvey—Obtain a general understanding of the problem by reading it quickly.
    - **Q**uestion—Find out what information is required in the problem.
    - **R**ead—Read the problem again to find information that is relevant to solving the problem.
    - **Q**uestion—Ask what operations must be performed and in which order to solve the problem.
    - **C**omplete—Do the computations necessary to get a solution.
    - **Q**uestion—Ask whether the answer is reasonable and the process complete.

WHO: Grade 3–Grade 5
WHEN: During the lesson
FOCAL POINT(S):   Measurement

- Have students memorize the formula for calculating the surface area of a cube with the following mnemonic:
    - **Length** × **Width**   **2 Little Worms**
    - **Length** × **Height**   **2 Little Holes**
    - **Width** × **Height**   (=) **2 Worm Holes**
    - _____   _____

(Martha Lamb, Curriculum Specialist, Middle Grades Literacy, Catawba County Schools)

WHO: Grade 6–Grade 8 (Advanced)
WHEN: During the lesson
FOCAL POINT(S):   Algebra

- Have students remember the mnemonic device, *Good Tomatoes Don't Lose Their Color,* to alert them that the process for solving a given absolute inequality (whether graphically or algebraically) will eventually yield a number–line solution. Therefore, the acronym, *Good Tomatoes Don't Lose Their Color* actually stands for if *Greater Than Diverges, Less Than Converges (GTDLTC)* (McNamara, 2006, p. 14).

WHO: Grade 6–Grade 8
WHEN: During the lesson
FOCAL POINT(S):   Algebra

- After being given examples for *solving linear systems by elimination,* have students create their own mnemonic devices to help them remember the steps in the process. For example, the acronym **MEST** would stand for the following:

M—**M**ultiply by a constant (if necessary).

E—**E**liminate a variable by adding like variables.

S—**S**ubstitute the value you get for one variable into either of the two original equations to get the value of the other variable.

T—**T**herefore, the solution to the system is (x, y).

Have students say **MEST.** Shout out each letter separately and wait for students to tell you what each stands for.

(Sonia Duggal, Woodstock Collegiate Institute, Thames Valley District School Board, London, Ontario, Canada)

WHO: Grade 6–Grade 8
WHEN: During the lesson
FOCAL POINT(S):   Geometry

• Use the following mnemonic device to help students remember the following three types of triangles:

Equilateral—The **E** in equilateral has three equal parts just as the triangle has three equal parts.

Scalene—The **S** in scalene has no equal sides, and the scalene triangle has no equal sides.

Isoceles—The **I** in isosceles has two equal sides and so does the Isoceles triangle.

(Mari Gates, Henry B. Burkland School, Middleborough, MA)

WHO: Grade 6–Grade 8 (Advanced)
WHEN: During the lesson
FOCAL POINT(S):   Geometry

• *Only Highlanders Are Happy On Adventures* is an acrostic that can be used to help students remember the following formulas:

$$\text{Sin O} = \frac{\text{Opposite}}{\text{Hypotenuse}}$$

$$\text{Cos O} = \frac{\text{Adjacent}}{\text{Hypotenuse}}$$

$$\text{Tan O} = \frac{\text{Opposite}}{\text{Adjacent}}$$

(Tony Duncan, Scott County High School, Huntsville, TN)

WHO: Grade 6–Grade 8 (Advanced)
WHEN: During the lesson
FOCAL POINT(S):   Geometry

- Have students remember the Indian Chief ***SOH–CAH–TOA*** when they are trying to recall the same trigonometric functions listed above (Ronis, 2006).

WHO: Kindergarten–Grade 8
WHEN: During the lesson
FOCAL POINT(S):   All

- Create original mnemonic devices (acronyms and acrostics) to assist your students in remembering mathematics concepts that need to be recalled. Have them recite the mnemonic device when recall of the concept is needed.

WHO: Grade 3–Grade 8
WHEN: During the lesson
FOCAL POINT(S):   All

- Have students create their own original mnemonic devices to help them recall mathematics formulas or concepts. Students will remember best what they choose to create themselves, especially if the mnemonic devices are humorous or novel.

REFLECTION AND APPLICATION

| How will I incorporate *mnemonic devices* into mathematics instruction with my students? |

Concept _____

Activity _____

_____

_____

Concept _____

Activity _____

_____

_____

Concept _____

Activity _____

_____

_____

Concept _____

Activity _____

_____

_____

Concept _____

Activity _____

_____

_____

Movement

WHAT: DEFINING THE STRATEGY

Several years ago, I was in the middle of teaching the *Worksheets Don't Grow Dendrites* class at Wiley High School in Wiley, Texas, when a football coach had an "Aha!" experience. I was teaching the part of the class where I talk about the power of procedural or *muscle* memory for the brain. I told the class that the skills learned when the body is moving are long remembered. This would include driving a car, riding a bike, or typing. The coach stood up in the middle of the class and exclaimed, "That explains it!" "That explains what?" I asked. The coach replied, "That explains why the football players that teachers tell me cannot retain their content in class can retain every play on the field." I thought this was an epiphany and commented that he was correct. While the players are running the plays on the field and placing them in procedural or muscle memory, in class they are immobile and not retaining much of the content being taught.

Such is the case with many math classes. While students could be role playing angles with their arms or doing the number-line hustle, they are sitting in their desks doing mathematics worksheets. When I teach model lessons in classrooms, I always integrate movement, especially for those bodily/kinesthetic learners who need it more than anyone. The entire class is so enamored of the lesson that they complain when it is over, and ask if I can return the following day. Of course I can't, because I am off to teach in another class. Movement is probably my favorite strategy since you can not only see how excited students get but how much they retain when it is used to teach any content.

WHY: THEORETICAL FRAMEWORK

Movement not only assists with reading, gets blood and glucose to the brain, changes the state or mood of the brain, and provides lots of fun during learning, it also assists with our strongest memory system—procedural memory (Sprenger, 2006b).

> Instructional situations that involve the use of movement necessitate more sensory input than do situations requiring only paper and pencil (Gregory & Parry, 2006).
>
> Repeat a movement often enough and that movement becomes a permanent memory (Sprenger, 2006b).
>
> Movement not only enhances learning and memory, it also causes neural connections to become stronger (Hannaford, 1995).
>
> The brain fuels itself on the oxygen in the blood, which is produced by physical activity (Sousa, 2006).
>
> A change of stimuli is crucial since the amount of time a student can focus is equivalent to the age of the student in minutes (DeFina, 2003).
>
> Man exists in a body/brain system, and man's intelligence is only one facet of that complicated system (Sylwester, 2000).
>
> When students integrate modern dance into their presentations, they are working on their kinesthetic and spatial skills (Mann, 1999).

HOW: INSTRUCTIONAL ACTIVITIES

WHO: Prekindergarten–Grade 1
WHEN: During the lesson
FOCAL POINT(S):   Algebra

• Help students recognize patterns by clapping or stomping out a pattern for the class and having the class repeat the pattern. The following is an example of one such pattern:

Clap, Clap, Clap

Stomp, Stomp

Clap, Clap, Clap

Stomp, Stomp

WHO: Kindergarten–Grade 2
WHEN: After the lesson
FOCAL POINT(S):   Number and Operations

• Have the entire class skip count aloud by 2s, 3s, 5s, 10s, 20s, etc. Add movement by having them clap or take turns jumping rope while skip counting.

WHO: Prekindergarten–Grade 1
WHEN: During the lesson
FOCAL POINT(S):   Number and Operations

• Teach appropriate directional concepts by having students move into different configurations. Knowing directions helps students when they begin to explore the concepts of place value and discuss the placement of digits in a number. For example, say the following:

- o *Students, please line up. Ben, please move **ahead of** Cheryl.*
- o *Joshua, stand **in front of** GeLinda.*
- o *Which student is standing **between** Mark and Jeremy?*
- o *Which student is standing **nearest** to me?*
- o *Who is standing **behind** Chris?*

WHO:　Grade 1–Grade 3
WHEN:　During the lesson
FOCAL POINT(S):　Number and Operations

- Create enough problems so that each student in the class can have an index card with a number that represents either a part of a problem or the solution. Give one student a plus sign, another a minus sign, and one an equal sign. Then begin stating the problems such as 3 + 2 = ?. The students with the 3, +, 2, and = cards get up, come to the front of the class and arrange themselves in the order of the problem. Then ask, "Who has the answer?" The student with the answer (5) should come to the front. Keep giving the class problems that involve different students.

To make this activity more challenging, give students only the answers to problems that you have created ahead of time. Read each problem aloud. Students who think they have the answer should stand and come to the front of the room. The class then verifies the answer.

WHO:　Grade 3–Grade 5
WHEN:　Before the lesson
FOCAL POINT(S):　Number and Operations

- **Division dance**—The four steps in long division can be taught with the *division dance*. The steps are divide, multiply, subtract, and bring down. Have students stand and make the following motions: As they say the word *divide*, have them make a division sign where the right arm forms the top of the sign and the left arm forms the indented part. Then have them say *multiply* and make an "X" with their arms to symbolize the multiplication sign. They should next say *subtract* and make a subtraction sign with their right arm, and finally say *bring down* and make a fist with their right hand and move their arm in a downward motion. This is called the *division dance,* and the four steps are repeated to the tune of any high-energy song played while students are making the signs.

WHO: Grade 3–Grade 5
WHEN: During the lesson
FOCAL POINT(S):   Number and Operations

  • Read a multiplication or a division problem to the class with an accompanying answer. Make sure that some of the answers are correct and some are incorrect. After the problem and its answer are read, have students stand if the answer is correct and remain seated if the answer is incorrect.

WHO: Grade 3–Grade 5
WHEN: During the lesson
FOCAL POINT(S):   Measurement

  • Have students work in cooperative groups taking turns to measure one another's arm length from the shoulder to the tip of the longest finger. They should then design a table containing the names of all students in the group and their arm measurements in feet and inches. The group can then figure who has the longest and shortest arm measurement. They may even want to convert the measurements from standard to metric units and find ratios between various measurements.

WHO: Grade 1–Grade 3
WHEN: After the lesson
FOCAL POINT(S):   Number and Operations, Measurement

  • Have students stand. Tell them to pretend that they are a magician's assistant and are about to be sawed in various spots on their bodies. They must show you where on their bodies they are about to be sawed. For example, tell students to pretend that they are being sawed in half, and to show you where the "magician" would saw them. Then have them show you the spots on their bodies that would be sawed if they were sawed into thirds, fourths, eighths, and so forth.

WHO: Grade 3–Grade 5
WHEN: During the lesson
FOCAL POINT(S):   Number and Operations

  • Write pairs of equivalent fractions (such as 1/2 and 5/10; 3/4 and 6/8) on separate index cards and pass them out randomly to students in the class. Be sure to have enough pairs so that every student has a card with a fraction on it. Play some fast-paced music and when you say go, have students rise from their seats, find the student with a fraction card equivalent to their own, and lock arms with that student. When all students have found a match, allow the class to determine if each pair of fractions is truly equivalent.

WHO: Grade 5–Grade 8
WHEN: Following the lesson
FOCAL POINT(S): Geometry

• Have students form a circle and use movement to review geometric terms. When the music starts, all students can move clockwise to show the circumference, another student can be selected to walk the radius, and another to walk the diameter of the circle. As you call out each term, students should demonstrate the term. Have students visualize this activity as they recall the definition of each term on a subsequent test.

WHO: Grade 5–Grade 8
WHEN: During the lesson
FOCAL POINT(S): Measurement, Geometry

• Have students move around the room and touch objects that correspond with geometric concepts. For example, have students take turns touching the following:

   o Five right angles
   o Three obtuse angles
   o One acute angle
   o Three sets of intersecting lines
   o Anything that is larger than one square meter
   o Anything that is smaller than one cubic meter
   o Five different rectangles
   o Something with a perimeter greater than one meter

After students come back to their seats, have three of them volunteer or select their names randomly out of a can to show what they touched.

(Darla R. Quinn, Grade 5, Johnstone Elementary School, Calgary, Alberta, Canada)

WHO: Grade 4–Grade 5
WHEN: During the lesson
FOCAL POINT(S): Measurement

• Have students use their bodies to demonstrate various distances in metric units. For example, when you say *centimeter,* students should show an approximate centimeter using the thumb and forefinger of one hand. For a *decimeter,* they should spread the fingers apart approximately 10 centimeters, or one decimeter. When the term *kilometer* is mentioned, students should drop their hands to their sides to show that a kilometer is so great that it cannot be shown with their body (Bender, 2005).

WHO: Grade 6–Grade 8
WHEN: During the lesson
FOCAL POINT(S):   Measurement

• Put customary systems of measurement and their metric counterpart on index cards. Pass them out randomly to students in the classroom. Put on some fast-paced music and have students walk around the room until they find their counterpart. When the music ends, have students share their matches with the class and have students determine if the match was made correctly.

WHO: Grade 6–Grade 8
WHEN: During the lesson
FOCAL POINT(S):   Number and Operations

• **Number-line hustle**—Draw a number line on the board. Explain the position of the positive and negative integers on the number line. Have small groups of students stand and tell them that they will be doing the *number-line hustle* by moving along the number line. (It may be cumbersome to do this with the whole class, depending on available space. I have tried it before and it works even if the students have to take small steps.) Have each student stand in a place in the room where they have space to move both left and right. They should all be facing in the same direction, preferably toward the number line on the board. You may use any appropriate disco music such as Van McCoy's *The Hustle*. Put on the music and then position yourself in front of the class. Lead the class in the movements necessary to solve the following problems:

Problem I: (+5 + –3 = ?) Have students move with you to the music five steps in a positive direction (+5). Then have them move three steps in a negative direction (–3). Ask them what number they landed on. The class should say (+2).

Ask the class what they would have to do to get back to zero. (They should say move two steps in a negative direction). Have them move back to zero.

Problem II: (–6 + +10 = ?) Starting at zero, have students move six steps in a negative direction (–6). Then have them move +10 steps in a positive direction. Ask the class, "What number are you on?" The class should say (+4).

Have the class sit down. Put the same problems that you danced out on the board so that students can see the connection between the concrete and the abstract. Then provide five additional problems for students to work either individually or in pairs. Give students the option of going to the back of the classroom and dancing along the number line to solve the additional problems while the music continues to play softly in the background.

WHO: Grade 6–Grade 8
WHEN: Before the lesson
FOCAL POINT(S):   Geometry

- Assist students in remembering the following formula for slope:

$$\textbf{Slope} = \frac{\textit{Rise}}{\textbf{Run}}$$

Have them **rise** from their chairs and **run** for the classroom door. Tell them that the formula does not work the other way around. It is impossible to **run** for the door without **rising** from their desks first.

WHO: Prekindergarten–Grade 8
WHEN: After the lesson
FOCAL POINT(S):   All

- Have students stand and form a circle. Tell students that they will be playing a *ball toss* game. The rules are as follows: Ask a question to all students that can be answered without paper and pencil. Allow wait time for all students to think of the answer. Then toss the ball to one student in the circle who will catch the ball and give you the answer to your question. If the answer is correct, that student can toss the ball to the next student in the circle. However, the new student cannot be standing in close proximity to the student with the ball. If the student misses the question, then the student must toss the ball back to you. You then toss the ball to another student in the circle. Students get one point for catching the ball and an additional two points for answering the question correctly. Be certain that all students have had an opportunity to catch the ball at least once before you conclude the game.

This ball toss game can be used with any grade level to review any math concept previously taught, such as the following: addition, subtraction, multiplication, or division facts, or meanings of math vocabulary words such as isosceles triangle, obtuse angle, or vertex.

REFLECTION AND APPLICATION

How will I incorporate *movement* into mathematics instruction with my students?

Concept _____

Activity _____

_____

_____

Concept _____

Activity _____

_____

_____

Concept _____

Activity _____

_____

_____

Concept _____

Activity _____

_____

_____

Concept _____

Activity _____

_____

_____

STRATEGY 11

Music, Rhythm, Rhyme, and Rap

WHAT: DEFINING THE STRATEGY

My daughter, Jessica, had nine years of piano lessons and is capable of sight reading music. In addition, she played the trombone in the high school band and sang in the chorus in both high school and college. She even auditioned for and made the chamber chorus in college. Jessica also scored very high on the mathematics section of the Scholastic Aptitude Test (SAT) and speaks such fluent German that she graduated from college with a major in that language. This may not be a coincidence.

Some research (Jensen, 2005; Sousa, 2006) suggests that the same spatial part of the brain that is activated when one is playing a musical instrument is also activated when one is speaking a second language or solving higher-level math problems. Some Japanese parents in my workshops have shared that they enroll their young children in Suzuki violin lessons realizing the possible correlation between their musical ability and their academic achievement. Just as there are rhythms and patterns in music, there are similar rhythms and patterns in math. This makes music a natural strategy for the teaching of mathematics.

WHY: THEORETICAL FRAMEWORK

Classical music by composers like Mozart and Beethoven stimulates beta waves in the brain and is appropriate for students to use when brainstorming or problem solving (Sprenger, 2006b).

Of all the content areas, mathematics appears to be the one most closely aligned with music. Music uses ratios, proportions, and fractions for tempo, patterns for notes or chords, counting for beats and rests, and geometry for placement of the fingers on a guitar (Sousa, 2006).

Fast music with 100 to 140 beats per minute can be energizing for the brain while calming music at 40 to 55 beats per minute can be relaxing (Jensen, 2005).

Certain musical selections from the Baroque period have fewer beats per minute and encourage the brain to calm down and relax (Sprenger, 2006b).

The rhythms, contrasts, and patterns of music help the brain encode new information, which is why students easily learn words to new songs (Webb & Webb, 1990; Jensen, 2005).

Rhythmic music can be played prior to the completion of a task that requires spatial reasoning, such as solving geometry problems or tackling a visual spatial problem. This type of music may enhance student performance (Jensen, 2005).

The mathematics scores of low socioeconomic students more than doubled for those who took music lessons compared to those who did not (Catterall, Chapleau, & Iwanga, 1999).

HOW: INSTRUCTIONAL ACTIVITIES

WHO: Kindergarten–Grade 1
WHEN: Before, during or after the lesson
FOCAL POINT(S):   Number and Operations

• Engage students in singing a variety of traditional "number" songs to help build their fluency with counting forward and counting backward. These songs include "One, Two, Buckle My Shoe" and "There Were 10 (insert an appropriate animal if desired) in a Bed."

WHO: Kindergarten–Grade 2
WHEN: Before, during or after the lesson
FOCAL POINT(S):   Number and Operations

• **Addition or subtraction rock**—Review addition and subtraction facts with students using either of the CDs *Addition/Subtraction Rock, Addition/Subtraction Rap* or *Addition/Subtraction Country.* Have students move to the beat as they recite their facts to the music. Each CD comes packaged with a cassette tape and book and can be found at a local teacher store or by logging on to www.rocknlearn.com.

WHO: Grade 2–Grade 3
WHEN: During the lesson
FOCAL POINT(S):   Number and Operations

• **Rounding**—Have students memorize the following rhyme as a rule for whether to round a number up or down.

Four or less, let it rest.

Five or more, raise the score.

WHO: Grade 3–Grade 5
WHEN: During the lesson
FOCAL POINT(S):   Number and Operations

- **Range**—Have students sing the following song to the tune of "Row, Row, Row Your Boat" to remember how to find the range of two numbers:

> *Range, range, range, you know*
>
> *All you have to do*
>
> *Subtract the greatest and the least*
>
> *It's easy as one, two.*

(Robin Kuketz, Grade 5, Henry B. Burkland School, Middleborough, MA)

WHO: Grade 3–Grade 5
WHEN: During the lesson
FOCAL POINT(S):   Number and Operations

- **Dividing fractions**—Have students memorize the following rhyme when dividing fractions:

> *Yours is not to question why.*
>
> *Just invert and multiply* (Ronis, 2006).

WHO: Grade 3–Grade 5
WHEN: Before the lesson
FOCAL POINT(S):   Number and Operations

- **Decimal place values**—Have students memorize the following poem as a rule for remembering how to read decimals.

> *Reading decimals is easy, you'll see*
>
> *They have two names like you and me.*
>
> *First, you say the name as if there were no dot*
>
> *Then you say the name of the last place value spot.*

Examples:	0.287	First name: two hundred eighty seven
		Last name: thousandths
	0.98	First name: ninety eight
		Last name: hundredths

(Robin Kuketz, Grade 5, Henry B. Burkland School, Middleborough, MA)

WHO: Grade 1–Grade 8
WHEN: Before the lesson
FOCAL POINT(S):   Number and Operations, Geometry, Measurement

- Help students make the connection between math and music by having the music teacher demonstrate how math uses ratios, proportions, and fractions for tempo, patterns for notes or chords, counting for beats and rests, and geometry for placement of the fingers on a guitar.

WHO: Grade 3–Grade 5
WHEN: Before the lesson
FOCAL POINT(S):   Number and Operations

- **Multiplication rap**—Review multiplication facts with students using either of the CDs *Multiplication Rap, Multiplication Rock,* or *Multiplication Country.* Have students move to the beat as they recite their tables to the music. Each CD comes packaged with a cassette tape and book and can be found at a local teacher store or by logging on to www.rocknlearn.com.

WHO: Prekindergarten–Grade 8
WHEN: After the lesson
FOCAL POINT(S):   All

- Put your creative talents to work! Write an original song, rhyme, or rap to symbolize your understanding of a math concept you have previously taught the class. Perform your creative effort for your students, and teach it to them so they can use the powerful effects of music to remember your content. They'll love you for it!

WHO: Grade 1–Grade 8
WHEN: After the lesson
FOCAL POINT(S):   All

- Following instruction in a major math concept, have students write an original song, rhyme, or rap to symbolize their understanding of the concept previously taught. Students can be assigned this task for homework, if class time does not permit. Then on the following day, all students can attend the *Talent Show* where volunteers pretend to be on *American Idol* and get up and perform their original effort for the class.

WHO: Grade 6–Grade 8 (Advanced)
WHEN: During the lesson
FOCAL POINT(S):   Algebra

- Assist students in memorizing the following three trigonometric math ratios for solving word problems about angles and sides for triangles. Have them do the movements and sing the following song to the tune of the "Macarena":

SOH CAH TOA

Sine is opposite over hypotenuse,

Cosine is adjacent over hypotenuse,

Tan is opposite over adjacent,

Ehhhhhhhh . . . . SOH CAH TOA!!

Repeat as many times as necessary and speed up the song as students master it.

(Sonia Duggal, Woodstock Collegiate Institute, Thames Valley District School Board, London, Ontario, Canada)

WHO: Grade 6–Grade 8
WHEN: During the lesson
FOCAL POINT(S):   Geometry

• Have students sing the following song to the tune of the "Hokey Pokey" to help them remember how to determine the slope of a line.

You put your y two in, You take your y one out,

You put your x two in, And then you yank your x one out.

Divide 'em in the middle, And drop an "m" at the start.

That's what slope is all about!

(Repeat)

(Sonia Duggal, Woodstock Collegiate Institute, Thames Valley District School Board, London, Onatrio, Canada)

WHO: Grade 3–Grade 5
WHEN: During the lesson
FOCAL POINT(S): Number and Operations

• When students have difficulty recognizing whether or not a fraction is in its simplest form, use the following rhymes:

1—One and Done

When there is a one in the numerator, you're done!

B—Back-to-Back Jack

Numbers in the fraction are in back-to-back order.

Ex. 3/4 , 5/6, 7/8, 11/12, and so forth

P—Prime is Slime

A prime number is in the denominator.

Ex. 3/5, 4/7, 5/11, and so forth

N—No Rule

Sometimes there is no rule to let you know if you are done, but here are a few hints:

o There is a prime number in the numerator.
o There is an even/odd combination or and odd/odd combination.
o The denominator is not a multiple of the numerator, such as 5/8, 4/9, or 7/12.

(Robin Kuketz, Grade 5, Henry B. Burkland School, Middleborough, MA)

WHO: Grade 6–Grade 8
WHEN: During the lesson
FOCAL POINT(S): Data Analysis and Probability

- *Mean, Median, Mode* Have students sing either of the following songs to the tune of *Three Blind Mice* to learn the characteristics of the mean, median, and mode.

(1)

Mean, Median, Mode

Mean, Median, Mode

Central tendency

Central tendency

The mean is the average of the lot.

The median's the one in the middle spot.

The mode is the one that you see a lot.

Mean, median, mode

(2)

Mean, median, mode

Mean, median, mode.

Went walking down the road.

Went walking down the road.

Mean adds and divides all the time.

Median stays in the middle of the line.

Mode happens most of the time.

Mean, median, mode.

(Theresa Esper, Grade 6, Harold L. Qualters Middle School, Mansfield, MA)

WHO: Grade 6–Grade 8
WHEN: During the lesson
FOCAL POINT(S): Geometry

- To assist students in finding the slope when given an equation in standard form, have them sing the following song to the tune of Disney's

Bare Necessities from *The Jungle Book:*

> *Look for the opposite of A over B*
> *The opposite of A over B*
> *Is the way to find the slope of a line*
> *Look for the opposite of A over B*
> *That's Mr. B's recipe*
> *To find the slope of a line in standard form*

(Bruce Wayne Basinger, Special Education Teacher, Cleveland High School, Reseda, CA)

WHO: Grade 6–Grade 8
WHEN: During the lesson
FOCAL POINT(S): Geometry

• To find the slope when given an equation in slope-intercept form, have students sing the following song to the tune of *Hit the Road Jack* by Ray Charles.

$$\boxed{A}x + \boxed{B}y = C \qquad \text{Standard Form}$$

$$\boxed{3}x + \boxed{5}y = C \qquad \text{Sample equation in standard form}$$

Slope = opposite of A over B

$$\boxed{\text{Slope} = m = -3/5}$$

> *What's the slope Jack*
> *It's the number*
> *Next to, next to,*
> *Next to X*
> *That's the slope Jack*
> *Of a line in slope-intercept form*

Finding slope of a slope-intercept equation

$$y = \boxed{m}x + b \qquad \text{Slope Intercept Formula}$$

$$y = \boxed{4}x + 7 \qquad \text{Example line in slope-Intercept form}$$

$$\boxed{\text{Slope} = m = 4}$$

(Bruce Wayne Basinger, Special Education Teacher, Cleveland High School, Reseda, CA)

WHO: Grade 3–Grade 8
WHEN: After the lesson
FOCAL POINT(S): All

• When trying to recall formulas to solve problems, have students create original songs. For example, have students sing the quadratic equation when needing to use it in problem-solving. It is sung to the tune of *Frere Jacques.*

$$X = \frac{-b \pm \sqrt{b^2 - 4ac}}{2a}$$

Negative B *(Repeat)*

Plus or Minus Square Root (Repeat)

B *Square Minus Four* AC *(Repeat)*

All Over Two A *(Repeat)*

(Tate 2003)

This same formula can be sung to the tune of *Pop Goes the Weasel.*

$$X = \frac{-b \pm \sqrt{b^2 - 4ac}}{2a}$$

X *Equals Negative* B

Plus or Minus Square Root

B *Square Minus Four* AC

All Over Two A

WHO: Grade 6–Grade 8
WHEN: Before the lesson
FOCAL POINT(S):   Geometry

• Use the following as a read aloud. Have students enjoy the clever rhyme and, in subsequent readings, have them interpret various sections of it.

Geometry Jingles

A crippled Scalene Triangle went limping with his cane

And met a jolly Circle upon a level plane.

"How do you do?" "And how are you?" they said like you and me

And then sat down together 'neath an old geome-tree.

And while they loitered in the shade they told the village news:

Right Triangle's baby crawls on its hypotenuse.

The Perpendiculars, they said, had quadruplets last night

And though it's much too soon to tell, they ought to be all right.

Obtuse Angle's home from school and with his parents stern.

He's not acute, he isn't right. He's just too dull to learn.

The Parallels are pleasant twins who never fuss or fret

Though they must act like strangers because they've never met.

Someone said that Polygon had eaten too much cake

And her indigestion's awful with so many sides to ache.

One Quadrilateral, they say, will always treat you fair.

He's quite a regular fellow and he signs his name "B. Square!"

His cousin Rhombus won't sit at all but slumps down in his chair

And all the teachers frown and say, "I wish you'd act like Square."

Scalene's cousin never limps but walks with greatest ease

Because his legs have equal length. His name's Isosceles.

A chap whose name is Trapezoid is shaped quite like a bin;

Two sides are level parallels and two sides just slope in.

The Point thinks he's important though he hasn't any size.

He knows he has position and position satisfies.

But should he move, he'll make a streak, a streak so very fine

It hasn't either depth or breadth. They call the streak a Line.

And when a Line gets restless and moves from here to there

It generates a Surface like a cube or sphere.

And if a Surface moves—Well! Well! Hang on and keep it steady

For the subject's getting solid and you're really not quite ready.

Soon Circle yawned and Scalene stretched and both sighed wearily

And in geome-tree's deep shade they slept like you and me.

REFLECTION AND APPLICATION

> **How will I incorporate _music, rhythm, rhyme,_ and _rap_ into mathematics instruction with my students?**

Concept _____

Activity _____

_____

_____

Concept _____

Activity _____

_____

_____

Concept _____

Activity _____

_____

_____

Concept _____

Activity _____

_____

_____

Concept _____

Activity _____

_____

_____

STRATEGY 12

Project-Based and Problem-Based Instruction

WHAT: DEFINING THE STRATEGY

The following is a true story. A fourth grade class was complaining about their dislike for math. *Why do we have to learn fractions? What difference does it make if we ever learn this stuff?* These questions could be heard reverberating throughout the classroom. Their very bright teacher decided to try the following project. On one Monday morning as students filed into the room, she told them that the day would be special—that she had heard their pleas and had decided that Monday would be "No Math Day!" The class cheered wildly realizing that, for once, they would not be involved in the usual math activities for one day in their lives.

The teacher began the usual language arts activities. However, while reading the entertaining story, *Flat Stanley,* it became apparent that there were shapes in the text. She immediately had students close their books and informed them that they could not continue their reading due to the fact that knowledge of shapes involved math and today was "No Math Day." The morning proceeded. One student noticed that the clock had been removed from the wall. When asked why, Mrs. Williams calmly stated, "Telling time involves mathematics and today is 'No Math Day.'"

As the morning progressed, lunch time was approaching. However, the lesson continued and no mention was made of a lunch period. When one child inquired as to why the class was not proceeding to the lunch room, Mrs. Williams stated that, without the clock, she had no idea when lunch would be. She also informed the students that she had also removed her watch and, therefore, also had no idea as to when the end of the school day would come.

Needless to day, the class was in total agreement that "No Math Day" needed to be canceled! Instead the teacher assigned a project to the class. The students were to select a profession of interest and list all the ways that the profession uses math. This project reinforced the need for and uses of mathematics so the students could answer "Why do we need math?" for themselves.

WHY: THEORETICAL FRAMEWORK

Problem-based instruction enables students to learn math content as they solve the same problems that people in the real world (architects, scientists, and engineers) have to solve (Ronis, 2006).

When a new math skill is viewed within the context of a problem, English language learners have opportunities to develop language skills through discussion (Coggins et al., 2007).

Parents should involve their children in real-world projects that involve reasoning or mathematical skills, such as planning for a birthday party, buying carpeting, or calculating a budget (Wall & Posamentier, 2006).

When students interact with other students in a group while solving problems, both cognitive (basic) and metacognitive (higher-order) thinking skills are stimulated (Posamentier & Jaye, 2005).

Project-based instruction provides the student with the intrinsic rewards of natural curiosity and a search for meaning (Ronis, 2006).

Educators should use such authentic tools as projects, discussions, and portfolios, in addition to paper-and-pencil tests to demonstrate students' comprehension of mathematics (Ronis, 2006).

Students who think out loud when solving problems have more awareness of what information is needed to solve the problem and, therefore, think more systematically (Posamentier & Jaye, 2005).

Skill-and-drill instruction may assist students with developing low-level mathematical skills but can destroy students' love of learning. Real-world inquiry methodologies challenge students to solve problems the way actual bankers, engineers, or architects do (Ronis, 2006).

The frontal and parietal lobes of the cerebrum in the brain are both involved when one is solving problems or processing mathematically (Sousa, 2001).

Students should have many opportunities to develop, *grapple with,* and solve difficult problems that necessitate a great deal of effort and should be provided with opportunities to reflect on their thought processes (National Council of Teachers of Mathematics, 2000).

HOW: INSTRUCTIONAL ACTIVITIES

WHO: Grade 1–Grade 8
WHEN: Before the lesson
FOCAL POINT(S):   All

• Have students create their own problems for other students to solve. When students enter the room, have one student's problem on the board for all students to solve as a *sponge activity*. The problem could provide a review of a problem-solving strategy already taught and could supply some needed practice. Students could earn extra points by solving one another's problems correctly, either individually or with a partner.

WHO: Grade 3–Grade 5
WHEN: During the lesson
FOCAL POINT(S): Number and Operations

• Create real-world problems incorporating the names of students in the class. Have students work individually, in pairs, or in small groups to solve the problems. For example, when teaching estimation, give students the following problem:

> Roderick, David, and Angela want to go to the movies. They have $35 between them. If a movie ticket cost $6.00, a Coke cost $2.50, and popcorn, $3.00, estimate whether all three classmates can afford to get into the movies with Coke and popcorn for each person.

Have students share the thought processes involved in estimating the answer to the aforementioned problem. Have students discern that there may be more than one way to figure out the answer.

WHO: Grade 3–Grade 5
WHEN: During the lesson
FOCAL POINT(S): Number and Operations and Measurement

• Have students measure elapsed time by taking on the following project. Have them pair with another student in class and relate the times that specific things happen in their day. For example, Karen, a student in class, shares the following schedule, which is placed on the board as an example:

- 6:15 a.m.   Gets out of bed
- 7:00 a.m.   Eats breakfast
- 7:30 a.m.   Catches school bus
- 7:55 a.m.   Arrives at school
- 12:15 p.m.   Eats lunch
- 3:30 p.m.   Arrives home from school
- 5:30 p.m.   Begins homework
- 7:00 p.m.   Eats dinner
- 9:00 p.m.   Goes to bed

Ask questions of the class such as: *How much time elapses from the time Karen gets out of bed until the time she arrives at school?*, or *How much time elapses from the time Karen eats lunch until the time she eats dinner?* Some students have a difficult time calculating elapsed time when the clock passes 12 a.m. or 12 p.m.

Once students understand the concept, their project becomes capturing the schedule of their partner and then creating elapsed-time problems that other students in class can solve.

WHO: Grade 3–Grade 5
WHEN: During the lesson
FOCAL POINT(S):   Number and Operations, Measurement

• Have students construct a class cookbook to apply their understanding of multiplying fractions. Have students find recipes for their favorite foods that have fractions of servings: for example, 2½ cups of flour, 2¼ cups of sugar, ¾ teaspoon of vanilla. Students then rewrite the recipe, cutting it in half and then doubling and tripling it. Students can choose one version of the recipe to make as a project for homework (Tate, 2003).

WHO: Grade 2–Grade 5
WHEN: During the lesson
FOCAL POINT(S):   Number and Operations, Measurement

• As a homework assignment, have each student write down the number of minutes it takes for them to do the following:
   o Brush their teeth
   o Comb their hair
   o Eat breakfast (if applicable)
   o Drink a glass of water
   o Travel to school
   o Complete homework
   o Eat dinner
   o Get ready for bed

Have them compare their recorded time to that of a partner. You can also help them figure out the class average for each activity.

WHO: Grade 3–Grade 8
WHEN: During the lesson
FOCAL POINT(S):   All

• Have students work in groups and follow the steps below when solving math problems:
   o Read the problem
   o Comprehend the problem
   o Analyze the problem
   o Plan an approach that can be used to solve the problem
   o Explore the approach to ascertain whether it will work
   o Use the plan to solve the problem
   o Verify the solution
   o Listen to and observe other students while solving the problem

(Posamentier & Jaye [2005])

These steps could be written in a graphic organizer and placed as a visual on the wall for students to follow when solving math problems.

WHO: Grade 2–Grade 8
WHEN: Before the lesson
FOCAL POINT(S):   All

- Empower students to learn from the process of discussing problem solving by modifying the "Problem of the Day" concept. Select a problem to begin the day, but instead of the focus being on solving the problem to get an answer, focus on discussing the problem. Here are some guidelines for "Problem Discussion of the Day". Address items such as:
    - ○ Characteristics of the problem
    - ○ Context of the problem
    - ○ Vocabulary in the problem
    - ○ Tools (tangible or conceptual) needed to solve the problem
    - ○ Type of solution required by the problem
    - ○ Possible ways of solving the problem
    - ○ How the problem relates to other problems seen before

WHO: Grade 3–Grade 8
WHEN: During the lesson
FOCAL POINT(S):   All

- Involve students in a project that requires them to collect and analyze data from a survey. Have students select a topic of interest to them and then determine how they will collect the data, what the sample size should be, who should be sampled, and what type of graph (circle, line, bar, etc.) would be best for depicting the data. Topics could include some similar to the following: *What is your favorite brand of toothpaste? Should students in this school wear uniforms? What foods should be served in the cafeteria most often? What is your favorite subject and why?*

WHO: Grade 6–Grade 8
WHEN: During the lesson
FOCAL POINT(S):   Number and Operations

- Engage students in interdisciplinary cooperative learning projects such as the following: The class forms into student teams of four to six. Each team selects one football, basketball, or baseball team to follow for 10 to 20 games of the regular season. Each team will choose the most valuable player of the team for the 10- to 20-week time period but must be ready to justify the choice using vital statistics as evidence. The team will plan and deliver a broadcast including a PowerPoint presentation during which they will perform a report analysis and interpretation of the stats. They will also submit journals in which they have tracked the team's statistics.

WHO: Grade 5–Grade 8
WHEN: During the lesson
FOCAL POINT(S):   Measurement, Data Analysis and Probability

- Involve students in an interdisciplinary project where they have to use tables and graphs to describe rate of growth. For example, divide students into groups and give each group a plant. Have each group measure the height of their plant in centimeters on the first day and then keep track of their plant's growth over the next 30 days. Show them how to construct a table and work together to record the changes in the plant's growth over time. Show them how to convert the table into a graph reflecting the same information.

WHO: Grade 3–Grade 8
WHEN: During the lesson
FOCAL POINT(S):   All

- When solving problems during class discussions, allow students to take turns sharing their ideas. Have them use sample sentence starters such as the following:
  - *I realized that . . .*
  - *I agree with your thinking and would like to add . . .*
  - *I don't understand what you meant when you said . . .*
  - *I solved the problem this way . . .*

WHO: Prekindergarten–Grade 8
WHEN: During the lesson
FOCAL POINT(S):   All

- Assist students with their problem-solving abilities by modeling the thought processes involved. Show them how to comprehend the problem and look back in the problem as they reflect on the solution and the process used to derive the solution. Students should record their processes to support discussion.

WHO: Grade 1–Grade 8
WHEN: Before the lesson
FOCAL POINT(S):   All

- When the math period begins or when students enter your math class, have a "Problem of the Day" on the board for students to solve either individually, with a partner, or in small groups. This problem should review skills learned in previous lessons and provide some much-needed practice for the brain. Relate the problem as closely as possible to that which students would encounter in the real world. For example:
  - The lunchroom staff needs to prepare lunches for the school today.
  - The students in the school are organized in grades kindergarten through fifth grade.

o Each grade level has three classes.
o Kindergarten through second grade have 17 students per class.
o Grades third through fifth have 24 students per class.
o If every student is eating lunch, how many lunches will lunch-room personnel need to prepare?

Have students use the steps for solving problems to solve the "Problem of the Day." (Look at the steps in the Activity referenced to Posamentier and Jaye in this chapter.)

REFLECTION AND APPLICATION

**How will I incorporate
project-based and *problem-based instruction*
into mathematics teaching with my students?**

Concept _____

Activity _____

_____

_____

Concept _____

Activity _____

_____

_____

Concept _____

Activity _____

_____

_____

Concept _____

Activity _____

_____

_____

Concept _____

Activity _____

_____

_____

Reciprocal Teaching and Cooperative Learning

WHAT: DEFINING THE STRATEGY

My daughter, Jennifer, is an excellent classroom teacher. She teaches the primary grades in the same school system that I worked in for 30 years. She is also an excellent presenter to other teachers. Therefore, I felt it was time for her to share her talents by presenting the same *Worksheets Don't Grow Dendrites* class that I present. For the past two summers, Jen has been touring with me and presenting parts of the class to my audiences. The parts that she doesn't teach during the day are taught to me in the hotel room at night. You see, you learn what you teach. In fact, to teach is to learn twice! This past summer, Jennifer presented five times to audiences across the country and received rave reviews. I had no doubt since she was well-prepared.

When students teach one another a math concept just taught to them by a teacher or when they work together in cooperative groups to discuss the fact there may be more than one way to solve a problem, learning is enhanced. No matter how well-presented the lesson, students' brains do not retain everything covered in one presentation. However, when students are able to share what they learned with a partner, and they glean from a partner what the partner has learned, comprehension and retention soar. After all, they speak one another's language and can often re-explain a concept in ways that are superior to the way that concept was taught initially.

WHY: THEORETICAL FRAMEWORK

Share what you know and feel memories grow (Sprenger, 2006b).

Individual students' abilities can be nurtured when those students belong to a community of learners who engage in peer tutoring and working collaboratively to make sense of mathematics (Posamentier & Jaye, 2005).

Children learn best when they have the opportunity to discuss ideas with their peers in a *nose to nose* and *toes to toes* interaction (Gregory & Parry, 2006).

Students who are unable to solve problems by themselves can often solve them when provided with temporary *scaffolds* or support that can be supplied by another student in class (Posamentier & Jaye, 2005).

Without the meta-cognitive process of group debriefing following a cooperative activity, there is only minimal improvement in the group's ability to use a specific collaborative or social skill (Gregory & Parry, 2006).

One of the 10 things that every child needs to build strong emotional intelligence is the ability to interact with others in their environment (McCormick Tribune Foundation, 2004).

Students' memory is strengthened when they are provided with opportunities to teach the entire class, partners, or small groups (Tileston, 2004).

Average and low achievers with or without learning disabilities showed greater achievement in classrooms where peer tutoring occurred than in those that did not have it (Posamentier & Jaye, 2005).

People remember 95% of what they are able to teach to someone else (Glasser, 1990).

People learn . . .

10% of what they read

20% of what they hear

30% of what they see

50% of what they both see and hear

70% of what they say as they talk

90% of what they say as they do a thing (Ekwall & Shanker [1988, p. 370])

HOW: INSTRUCTIONAL ACTIVITIES

WHO: Prekindergarten–Grade 8
WHEN: During or after the lesson
FOCAL POINT(S):   All

• Have each student select a *close partner*, a fellow student who sits close by in class so they can talk with this person, whenever necessary, and not have to get out of their seat. Stop periodically during a math lesson and have students re-teach a chunk of information just taught, brainstorm an idea, or review content prior to a test. Close partners can also re-explain a math concept that might not be clear or easily understood by their partner (Tate, 2006).

WHO: Grade 2–Grade 8
WHEN: During the lesson
FOCAL POINT(S):   All

• Have students work together in cooperative groups, or *families*, of four to six students. They may be seated already in groups or taught to arrange their desks into groups for a cooperative learning activity and to put them back once the activity is over. Groups should be of mixed ability levels to capitalize on the various multiple intelligences of students.

Give each group the same math problem to solve. Have them discuss the thought processes involved in solving the problem and reach a consensus as to the correct answer. Once the group agrees on an answer, have each person in the group sign the paper the answer is written on, verifying that they agree with the answer and, if randomly called upon, could explain how the solution was derived to the entire class. This individual accountability helps to ensure that one person does not do all the work while other students watch their efforts.

WHO: Grade 1–Grade 8
WHEN: During the lesson
FOCAL POINT(S):   All

• When students have difficulty working together as a cooperative group, you may want to teach some social skills necessary for effective functioning. For example, construct a *T-chart* similar to the one below where each social skill is considered from two perspectives: *what it looks like* and *what it sounds like*. Social skills could include the following: paying undivided attention, encouraging one another, or critiquing ideas and not peers.

Encouraging

Looks Like	Sounds Like
Heads nodding	Way to go!
One person speaking	Good job!
Smiles	Good idea!
Eye contact	What do you think?

Observe each group, making a tally mark on a sheet each time the social skill is practiced by any student in the group. Provide feedback to the class during a debriefing following the cooperative activity. You may also assign a student in each group to fulfill the function of a *process observer* who collects the data for the group.

WHO: Grade 1–Grade 8
WHEN: During the lesson
FOCAL POINT(S):   All

- Another way to help ensure individual accountability is to assign group roles for students to fulfill during the cooperative learning activity. Some of the following roles can be assigned:

  1. **Facilitator**—Ensures that the group stays on task and completes the assigned activity

  2. **Scribe**—Writes down anything the group has to submit in writing

  3. **Timekeeper**—Tells the group when half the time is over and when there is one minute remaining

  4. **Reporter**—Gives an oral presentation to the class regarding the results of the group's work

  5. **Materials manager**—Collects any materials or other resources that the group needs to complete the task

  6. **Process observer**—Provides feedback to the group on how well they practiced their social skills during the cooperative learning activity

WHO: Grade 1–Grade 8
WHEN: Before the lesson
FOCAL POINT(S):   All

- Have students draw the face of a clock. Have them write the numbers 12, 3, 6, and 9 in the appropriate places on the clock. Have students draw one line near each number. This clock becomes their vehicle for making appointments with their peers in class. Put on some fast-paced music and have students walk around the classroom and make appointments with four different students in class. They should write the name of the person they made each appointment with on the line next to the number so that later in the class, when it is time to meet with their appointment, they can remember with whom they made the appointment.

  Stop periodically throughout the period or the day and have students keep their appointments. Appointments can be used to re-teach a math concept previously taught or to discuss an open-ended question pertaining to the lesson.

WHO: Grade 2–Grade 8
WHEN: Before the lesson
FOCAL POINT(S):   All

- For variety, have students make appointments, like those in the aforementioned activity, but have them use the cycle of the seasons of the year, rather than the face of the clock. The cycle is pictured below. Follow the same directions as those listed in the previous activity.

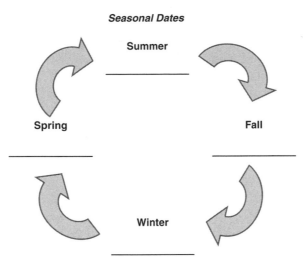

Seasonal Dates

WHO: Grade 3–Grade 8
WHEN: After the lesson
FOCAL POINT(S): All

• Find an article with graphics in the newspaper or on the Internet that reinforces a math skill previously taught, such as creating bar graphs or applying the concept of percentage; then make a copy for all students. Have students work in cooperative groups to study the data shown and make inferences regarding the real-life information. Have students make up questions for their classmates to answer regarding the data shown. Questions could include the following:

   o How is the math skill reflected in the article?
   o Can you make sense of how the math skill is applied in the article?
   o How is the math skill applied in the article different or the same when compared to how the math skill was taught in class?

WHO: Grade 1–Grade 8
WHEN: During the lesson
FOCAL POINT(S): All

• When students are solving math problems, have them use the *think, pair, share* technique. Students first individually *think* how they would solve the problem, then they *pair* with another student, and *share* their thought processes and answer to the problem. Both students should reach consensus as to the correct answer. If their original answers differ, the discussion involved in convincing their partner that they are correct is invaluable to the learning experience.

WHO: Grade 2–Grade 8
WHEN: During the lesson
FOCAL POINT(S): All

• Place students in cooperative groups. Have them participate in an activity called *Jigsaw. Jigsaw's* name is derived from the fact that each student has only one piece of the puzzle, and it will take all students to make a whole. Each student in the cooperative group is accountable for teaching one section of a math chapter or explaining one part of a multi-step problem to the entire group. The procedure is as follows:

1. Give students time to prepare their part individually either in class or for homework.

2. Have them confer with a student in another group who has the same part they do to get and give ideas prior to teaching their original group.

3. Give students a required number of minutes to teach their part to their original cooperative group. Individuals in each group start and stop teaching at the same time. If they finish before time is called, students can quiz group members for understanding.

4. Conduct a whole class review that outlines the pertinent points that should have been made during each student's instruction. In this way, the entire class gets to hear the content at least twice, once from their group member and once from the teacher.

WHO: Grade 1–Grade 8
WHEN: During the lesson
FOCAL POINT(S):   All

• While checking a homework or in-class assignment, have students take their paper and pencil in hand and walk around the room to fast-paced music. Every time you stop the music, have each student pair with another student standing close by and *give* one answer to their partner and *get* one answer from their partner.

WHO: Grade 1–Grade 8
WHEN: After the lesson
FOCAL POINT(S):   All

• Have students generate original questions to ask of their peers following a math lesson. Students appear more motivated and lessons become more purposeful when students are answering their own questions rather than those provided by the teacher or the textbook.

WHO: Prekindergarten–Grade 8
WHEN: Before, during or after the lesson
FOCAL POINT(S):   All

• When students are working with peers in small groups or talking to a partner, it is often difficult to get their attention. Create a signal and use it whenever you need students to pay attention to you. The signal can be a chime, a raised hand, a chant, a bell, or anything soothing that would not be abrasive to the brains of your students.

REFLECTION AND APPLICATION

> ## How will I incorporate *reciprocal teaching* and *cooperative learning* into mathematics instruction with my students?

Concept _____

Activity _____

_____

_____

Concept _____

Activity _____

_____

_____

Concept _____

Activity _____

_____

_____

Concept _____

Activity _____

_____

_____

Concept _____

Activity _____

_____

_____

Role Plays, Drama, Pantomimes, and Charades

WHAT: DEFINING THE STRATEGY

My son, Christopher, applied for a customer service position with a major company. Since Chris would have to deal directly with customers, the interviewer engaged him in a role play as a part of the interviewing process. She pretended that she was totally dissatisfied with the way her contract had been handled and had decided to take her business elsewhere. Chris' job was to convince her to stay with the company. It's a good thing that I equipped him with appropriate interpersonal skills, because he completed the role play successfully and was hired for the position.

While all of the 20 brain-compatible strategies correlate with long-term retention, role play is probably the strategy that places students' brains closer to the reason they exist in the first place, to survive in the real-life world. When students actually act out a math word problem, it becomes real and easier to understand. Have students come to the front of the room and become the major components of the problem. Not only will this serve as a visual for the remainder of the class, but when the time for the test comes, all students will be visualizing their classmates as they attempt to recall the concept taught.

WHY: THEORETICAL FRAMEWORK

Role play can be used to reinforce learning in a mathematics class. Examples would include having students make change, take measurements, or discuss algebraic equations (Sprenger, 2006a).

As students involve their bodies in the comprehension of concepts and ideas through role plays and skits, they are able to understand the material in a new way (Sprenger, 2006a).

Role plays use visual, spatial, linguistic, and bodily modalities and, therefore, not only access students' emotions but enable students to comprehend at much deeper levels than a lecture would (Gregory & Parry, 2006).

It can be a very engaging and highly effective activity to have a group of students act out or role-play a word problem (Bender, 2005).

Role play and pretend play enable children to experience empathy and determine what course of action is necessary in any given social situation (McCormick Tribune Foundation, 2004).

Role play makes learning more enjoyable, results in less stress from evaluation, and gives learners more creativity and choice (Jensen, 2000).

Simulations increase meaning, facilitate the transfer of knowledge, and are highly motivating (Wolfe, 2001).

HOW: INSTRUCTIONAL ACTIVITIES

WHO: Kindergarten–Grade 1
WHEN: During the lesson
FOCAL POINT(S):   Number and Operations

• Have students come to the front of the room and role-play (or act out) specific word problems. For example, for the following word problem, have six students come to the front of the class and stand in a row. Then say the following: *Six students are standing in a row. Two sit down.* (Have two students go back to their seats.) *How many students are left?* The class then counts the number of remaining students. Create other word problems by having students come to the front of the class and role-play the problem.

WHO: Prekindergarten–Grade 2
WHEN: After the lesson
FOCAL POINT(S):   Number and Operations

• Set up a classroom store with a cash register or check-out device. Place items in the store that sell for the increments of money previously studied, such as a penny, nickel, dime, quarter, or dollar. Place play money in the store and have students take turns buying and selling the items. Students can practice the skills of counting, determining the coins necessary to purchase specific items, making change, and so forth.

WHO: Grade 1–Grade 3
WHEN: During the lesson
FOCAL POINT(S): Number and Operations, Measurement

• Have 12 students at a time stand in a circle forming a clock. Have another student stand in the center of the circle. Give that student one short hand and one long hand to hold. Explain to students how the "12" and "6" positions are directly across from one another and so are the "3" and "9" positions. To demonstrate the hour numbers on the clock, have each of the 12 students say their name aloud, such as one o'clock, two o'clock, and so forth. To show the minutes on the clock have them count by fives beginning with the one position, such as five, 10, 15, 20, and so forth.

As you name a specific time, have the students who represent numbers involved in that time turn around in a circle. For example, if the time is 3:45, then the student standing in the "3" position and the one standing at the "9" position would turn around in a circle. Then the student standing in the center of the circle would use their short and long hands to show the actual time of 3:45. Explain to students that since 3:45 is past the "3" on the clock, the short hand would not be pointing to the student standing on the "3," but past that student. The remaining students who are not a part of the clock can take turns standing in the center and showing designated times with their short and long hands.

WHO: Kindergarten–Grade 8
WHEN: During the lesson
FOCAL POINT(S): All

• To more clearly understand the steps in a multistep word problem, have students take turns getting up and acting out each step of the word problem. This role play will work with a large majority of math problems and can go a long way in helping students who need a visual depiction of the problem.

WHO: Grade 3–Grade 5
WHEN: After the lesson
FOCAL POINT(S): Measurement

• Have students role play or pretend they are seamstresses or tailors. Have them work in pairs and take specific measurements of their partners such as their height, circumference of the head, and length from shoulder to tip of finger. Make a chart of the classes' measurements and have students convert them from standard to metric measurements.

WHO: Grade 5–Grade 7
WHEN: During the lesson
FOCAL POINT(S): Geometry

• Teach geometric terms ensuring that students understand their meanings, terms such as *line, line segment, ray, right angle, obtuse angle,* and *acute angle.* Show students an action for each definition. Then have students stand up beside their desks and role-play the definitions that you just demonstrated. For example, to demonstrate the term *line,* have students point both arms out to their sides and point their fingers to indicate that a line has no end points. To demonstrate *line segment,* have them point their arms straight out but ball their fingers into fists to demonstrate that a *line segment* has two end points. Have them demonstrate a *ray* by pointing the arms out and making the left fingers into a fist while pointing the right fingers out straight.

Students can role play *angles* by extending both arms to simulate *right, obtuse,* and *acute angles.* Involve students in a game by having them use their arms to make the terms as you randomly say the terms. This game is a lot of fun while putting the terms into procedural or muscle memory.

WHO: Grade 3–Grade 8
WHEN: After the lesson
FOCAL POINT(S):   All

•   Have students work in cooperative groups to prepare a television news broadcast related to a concept previously taught in math class. By the time students complete this interdisciplinary project, they will have written copy for the news broadcast summarizing the math concept taught; decided which visuals or graphics they need in the broadcast (including a PowerPoint presentation, if necessary); and selected a member of the group to deliver the broadcast on air. The class will use a rubric they helped to develop to vote on the most effective presentation.

WHO: Grade 1–Grade 8
WHEN: After the lesson
FOCAL POINT(S):   All

•   Review mathematics vocabulary by playing *Charades.* Write the words on separate 3″ × 5″ index cards. Have students take turns coming to the front of the room, selecting a card at random and acting out or role-playing the definition of the selected word. The student cannot speak but must use only gestures to get students to name the word. The first student to guess which vocabulary word is being acted out wins a point for guessing the word. The student with the most points at the culmination of the game is the winner.

WHO: Grade 6–Grade 8
WHEN: During the lesson
FOCAL POINT(S):   Geometry

•   Engage students in *Slope Aerobics* by having them stand and use their arms to role-play the following geometric terms:
   o   Positive slope—Students show a positive slope by spreading the arms out with the right arm up and the left arm down.

o Negative slope—Students show a negative slope by spreading the arms out with the left arm up and the right arm down.

o Zero slope—Students spread both arms straight out to their sides.

o Undefined slope—Students place the right arm straight in the air and the left arm straight down toward the floor to make a straight line.

Call out different slopes at a progressively faster pace as students make the slopes with their arms. Turn this activity into a game by having students sit down if they make a mistake with their arms. Give a prize for the last student standing. Watch students moving their arms to help them remember as they take tests.

(Theresa Galea, H.B. Secondary School, Thames Valley District School Board, London, Ontario, Canada)

WHO: Grade 1–Grade 8
WHEN: After the lesson
FOCAL POINT(S): All

• Have students take turns role-playing as if they are you, the math teacher. Have them volunteer to come to the front of the room and pretend, as the teacher, to re-teach the lesson previously taught. Give each student a maximum of three minutes. This activity will give you an idea of which concepts have been understood by your students and which need re-teaching. Remember that you learn what you teach and that most brains need to hear something a minimum of three times before the information actually sticks.

REFLECTION AND APPLICATION

How will I incorporate *role plays,*
drama, pantomimes, and *charades* into
mathematics instruction with my students?

Concept _____

Activity _____

_____

_____

Concept _____

Activity _____

_____

_____

Concept _____

Activity _____

_____

_____

Concept _____

Activity _____

_____

_____

Concept _____

Activity _____

_____

_____

STRATEGY 15

Storytelling

WHAT: DEFINING THE STRATEGY

In the children's book, *The Doorbell Rang,* by Pat Hutchins, a mother is baking cookies in the kitchen. Every time a new group of neighborhood children ring the doorbell and are invited in, the baked cookies have to be divided equally, meaning fewer and fewer cookies per person for the ever-increasing brood. Students are therefore introduced to the concept of division in a relevant, meaningful way. In the book, *Counting on Frank,* by Rod Clement, a boy named Frank figures out how to use unconventional units to measure various areas of his house. The illustrations are funny and the story is mathematically thought-provoking in a humorous way.

These are just two of the numerous examples of quality children's literature that can be used to introduce, teach, or review mathematical concepts. This literature takes advantage of the power of story to engage the frontal lobes of students' brains and enable them to think at higher cognitive levels. When teachers also tell stories that teach or reinforce points made during a math lesson, the point being made is remembered. Watch a keynote speaker or a minister tell a story during a speech or sermon and notice how everyone not only listens but more easily recalls the content of the speech or sermon.

WHY: THEORETICAL FRAMEWORK

Teacher-created stories can help students understand number operations in a variety of contexts (National Council of Teachers of Mathematics, 2000).

After a period of intense learning, storytelling enables the brain to relax and facilitates the retention of newly acquired material (Jensen, 2000).

The conflict or plot of a story can be addressed through emotional memory (Sprenger, 1999).

Story problems in which real objects are distributed in equal shares help prekindergarten through Grade 2 students understand the concepts of division and multiplication (National Council of Teachers of Mathematics, 2000).

Students' ability to listen and reason is improved during storytelling because they use the auditory modality with the frontal lobes of the brain to follow the story's plot (Storm, 1999).

Information is bound in our memories to the scripts that stories can provide (Markowitz & Jensen, 1999).

Storytelling ties information together and assists natural memory and, therefore, is a natural process for organizing information in the brain (Caine & Caine, 1994).

HOW: INSTRUCTIONAL ACTIVITIES

WHO: Kindergarten–Grade 8
WHEN: Before the lesson
FOCAL POINT(S): All

• Read aloud to the class any of the appropriate books listed in Resource A: Have You Read Any Math Lately? starting on page 147. Suggested books include *Skittles Riddles Math* by Barbara McGrath or *How Big Is a Foot?* by Rolf Myller. The first time you read the books aloud, do so for the enjoyment of the literature. Then revisit the book at another time and select a math skill or strategy from the story. The context of the story will help students remember the skill being taught.

WHO: Prekindergarten–Grade 3
WHEN: Before the lesson
FOCAL POINT(S): Number and Operations

• Find children's literature that contains math concepts. As you read a story aloud, have students listen for the concepts in the story. For example, in Hutchins' *The Doorbell Rang*, every time the doorbell rings, more and more children arrive and 12 baked cookies have to be divided equally. This story is wonderful for introducing the concept of division.

WHO: Kindergarten–Grade 8
WHEN: Before the lesson
FOCAL POINT(S): All

• The text of *Math and Children's Literature* by Carol Hurst provides numerous examples of stories that can be used to teach various math principles. Refer to the index to find children's books that will teach or reinforce a focal point being taught.

WHO: Grade 3–Grade 8
WHEN: During the lesson
FOCAL POINT(S): All

• After students have used manipulatives to develop and practice a concept, give them an equation and ask them to work individually or with a partner or group to create a real-life story explaining the equation.

WHO: Grade 3–Grade 5
WHEN: During the lesson
FOCAL POINT(S): Number and Operations

• When changing a fraction to a decimal, some students have difficulty figuring out which number to divide by. The following story often helps:

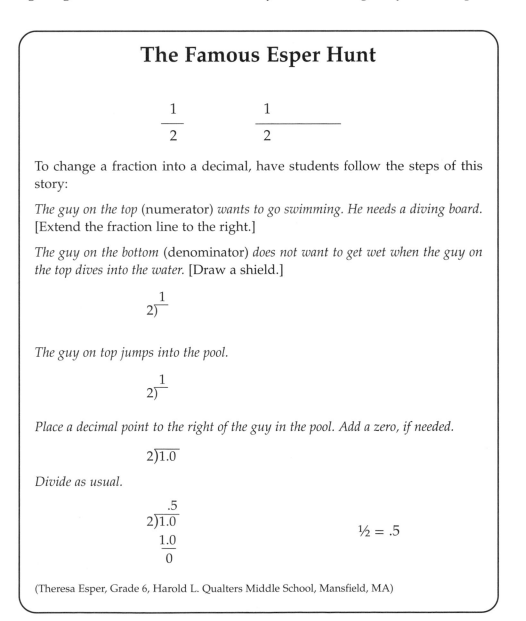

The Famous Esper Hunt

$$\frac{1}{2} \qquad \frac{1}{2}$$

To change a fraction into a decimal, have students follow the steps of this story:

The guy on the top (numerator) *wants to go swimming. He needs a diving board.* [Extend the fraction line to the right.]

The guy on the bottom (denominator) *does not want to get wet when the guy on the top dives into the water.* [Draw a shield.]

$$2\overline{)\tfrac{1}{}}$$

The guy on top jumps into the pool.

$$2\overline{)\tfrac{1}{}}$$

Place a decimal point to the right of the guy in the pool. Add a zero, if needed.

$$2\overline{)1.0}$$

Divide as usual.

$$\begin{array}{r} .5 \\ 2\overline{)1.0} \\ \underline{1.0} \\ 0 \end{array} \qquad \tfrac{1}{2} = .5$$

(Theresa Esper, Grade 6, Harold L. Qualters Middle School, Mansfield, MA)

WHO: Grade 6–Grade 8
WHEN: Before the lesson
FOCAL POINT(S): Algebra

• **The story of the algebraic equation**—Tell students the following story to teach them that when solving algebraic equations, the sign of the integer changes when the integer changes sides. Have seven students in the class hold up the seven parts of the following equation and act it out as you read this story.

$$3y + 10 = 2y + 18$$

[On the other side of the card for **10,** write **–10.** On the other side of the card for **2y,** write a **–2y.**]

Once upon a time there were two families who lived on either side of a busy street called *equal street.* Each family had two children (one teenage daughter and one younger son). One day the teenage daughters **3y** and **2y** made a date to go to the mall. However, there was a problem. Each daughter had been asked to babysit a younger brother. **3y** had to babysit **10** and **2y** had to babysit **18.**

The girls desperately wanted to get together to go to the mall. Therefore, daughter **3y** suggested that she send younger brother **10,** to cross *equal street* so that he could play with her friend's brother, **18.** Now there was one peculiar thing about this particular town. Whenever anyone crossed *equal street,* they had to turn around and cross it backward. [Turn the **10** card over to **–10.**] So younger brother **10** turned backward and crossed *equal street.* The two boys were very happy because now they could play together.

There was only one problem. In order to really be all alone with no one to bother them, the brothers had to get rid of big sister **2y.** Now that was alright with big sister **2y** since she wanted to go to the mall with her friend **3y** anyway. So she said goodbye to her brother and crossed *equal street* backward, of course, and she and her friend **3y** could be all alone to proceed to the mall. [Turn the **2y** card over to **–2y.**] The girls had a wonderful time and so did the boys.

Once the girls returned from the mall, they were in a world of trouble since their parents had told them over and over again never to leave their younger brothers unattended. But you know as well as I do that for generations older sisters have always left younger brother unattended. That's just the way the story goes (Tate, 2003).

WHO: Kindergarten–Grade 8
WHEN: After the lesson
FOCAL POINT(S): All

• Once students have understood a math concept, such as regrouping, create a story that incorporates characters in the real world to help them comprehend the concept and make the learning more relevant.

WHO: Grade 6–Grade 8
WHEN: Before the lesson
FOCAL POINT(S):   All

- To help students understand the multiplication of positive and negative integers, tell the following story:

  o When a good person (+) moves to (+) a good neighborhood, that is a good thing (+). Mathematically, **(+) × (+) = (+)**

    (A positive [×] a positive = a positive)

  o When a good person (+) leaves (–) a good neighborhood, that is a bad thing (–). Mathematically, **(+) × (–) = (–)**

    (A positive [×] a negative = a negative)

  o When a bad person (–) moves to (+) a good neighborhood, that is a bad thing (–). Mathematically, **(–) × (+) = (–)**

    (A negative [×] a positive = a negative)

  o When a bad person (–) leaves (–) a good neighborhood, that is a good thing (+). Mathematically, **(–) × (–) = (+)**

    (A negative [×] a negative = a positive)

(Payne, R. K. [2001])

WHO: Grade 3–Grade 8
WHEN: After the lesson
FOCAL POINT(S):   All

- Once students understand a math concept, have them work individually or in small groups to create an original story that incorporates characters in the real world to help them comprehend the concept and make the learning more relevant.

WHO: Grade 3–Grade 8
WHEN: During the lesson
FOCAL POINT(S):   Number and Operations

- Have students create a comic book story line with a fictional superhero who carries out the functions of a mathematical concept. Math superhero names could include *Addition Man, Polygon Woman,* or *Pythagorean Boy.* Students should create the story and draw the accompanying cartoons with dialogue for the character's speech bubbles.

REFLECTION AND APPLICATION

How will I incorporate *storytelling* into mathematics instruction with my students?

Concept _____

Activity _____

_____

_____

Concept _____

Activity _____

_____

_____

Concept _____

Activity _____

_____

_____

Concept _____

Activity _____

_____

_____

Concept _____

Activity _____

_____

_____

STRATEGY 16

Technology

WHAT: DEFINING THE STRATEGY

I was teaching in Plymouth, Massachusetts, when one of my clients, Nicola Micozzi, science coordinator and a technological *Nostradamas*, introduced me to a concept called *Second Life.* This unique Web site enables one to create a technological replica of oneself, called an *avatar.* This replica can be a human being with all the self-selected features that one wishes one actually possessed or an animal such as a squirrel or bird. The *avatar* can then be teleported around the Web site along with the nine million other *avatars* who are already on the site. Before Nicola had finished, I witnessed his *avatar*, a handsome Italian gentleman, flying to a seat in a lecture hall where he could communicate either verbally or in writing with other *avatars* from around the world who were all attending the same lecture.

Advances in modern technology have enabled us to accomplish tasks that were once thought impossible. Math is no exception. Calculators and computer software programs enable students to apply formulas and deal with amounts of data so large that in the past it would have taken days or weeks to problem-solve what now takes only seconds.

WHY: THEORETICAL FRAMEWORK

Higher achievement and greater understanding in math is achieved when technology is used for non-routine applications and not for routine calculations (Sousa, 2007).

When the inquiry-based learning model is paired with dynamic geometric software programs, students are able to discover relationships, make hypotheses, and defend assumptions (Posamentier & Jaye, 2005).

When calculators are used to explore and problem-solve in math, students view them as tools that can only enhance the learning (Guerrero, Walker, & Dugdale, 2004).

Technology can assist all students in developing number sense, but it is particularly helpful for students with special needs (National Council of Teachers of Mathematics, 2000).

> Calculators are useful for developing and reinforcing place-value concepts (National Council of Teachers of Mathematics, 2000).
>
> Students can perform complicated computations and work on real-life problems with electronic computation technologies (National Council of Teachers of Mathematics, 2000).
>
> Having students use technology to turn, deform, and shrink two- and three-dimensional objects helps them to develop the visualization skills necessary for comprehending geometric concepts in the early years (National Council of Teachers of Mathematics, 2000).

HOW: INSTRUCTIONAL ACTIVITIES

WHO: Grade 3–Grade 8
WHEN: After the lesson
FOCAL POINT(S):   All

• Have student access the Internet to locate information they can use in posing and solving problems of interest, such as: *What is causing increases in the intensity of certain types of weather across the country?*

• Students can also visit Ask Dr. Math at http://mathforum.org/dr.math/and pose their own inquiries about math.

WHO: Grade 6–Grade 8
WHEN: During or after the lesson
FOCAL POINT(S):   All

• Have students use graphing or scientific calculators as well as computers to enable students to solve problems with very large numbers or large quantities of data.

WHO: Grade 6–Grade 8
WHEN: During or after the lesson
FOCAL POINT(S):   All

• Have students use the *TI-73* or *TI-83/84 Plus* graphing calculators to solve equations.

WHO: Grade 3–Grade 8
WHEN: After the lesson
FOCAL POINT(S):   All

• Have students search the Internet for jobs in the real world that require the use of a math focal point like algebra. Since the purpose of the brain is to survive in the real world, then students often ask teachers why they have to learn a specific skill or concept and what that concept has to do with their survival in the real world. By engaging students in this conversation, that question might not have to be asked.

WHO: Grade 1–Grade 3
WHEN: After the lesson
FOCAL POINT(S): Geometry

- Have students use a drawing program on the computer to draw several duplicates of the same object, enlarge the object to various sizes, and then put the objects in order from smallest to largest.

WHO: Grade 1–Grade 5
WHEN: After the lesson
FOCAL POINT(S): Geometry

- Show students computer programs that extrude squares to form boxes and circles to form spheres. Following their use of the program, have students then draw two-dimensional figures and match them to their three-dimensional counterparts.

WHO: Grade 3–Grade 8
WHEN: After the lesson
FOCAL POINT(S): All

- **PowerPoint math**—Have students use a graphic organizer (sequencing map, flow chart) to frame the steps involved in solving a math equation or to explain a math term or concept. Have them decide which manipulatives or mediums they should use in photographs to replicate the information on the graphic organizer. Then have students take a photo for each segment of the information on the graphic organizer and upload the photos to the computer. Once the photographs are uploaded, have students insert them in PowerPoint frames and share their original creations with the class.

(Tomiko T. Smalls, Grade 2 Teacher, Mossy Creek Elementary School, North Augusta, SC)

WHO: Grade 3–Grade 8
WHEN: After the lesson
FOCAL POINT(S): All

- After students create an original math song, rhyme, or a rap, as suggested in Chapter 11: *Music,* have students follow the necessary directions to post their song on iTunes.

WHO: Grade 3–Grade 8
WHEN: After the lesson
FOCAL POINT(S): All

- After students create an original song, rhyme, or rap, as suggested in Chapter 11: *Music,* or create a visual such as the original commercial in Chapter 19: *Work Study,* have them post their creative effort on YouTube.

WHO: Grade 6–Grade 8 (Advanced)
WHEN: After the lesson
FOCAL POINT(S):   All

- **Picture graphing project**—To apply what they have learned about equations and conic sections, engage students in a project to create a program for the *TI-82* graphing calculator that will draw a picture. After drawing a picture on graph paper, students should then write the equations needed to draw the picture using the calculator. Each equation should be solved for *y*. Refer students to the user's guide to learn how to program the equations into their calculators.

Here are the student directions for this cooperative learning project:

- ○ Create a program using a graphing calculator that will draw a picture using the equations you have learned so far in this class.
- ○ You must use a minimum of 10 non-vertical line equations.
- ○ Two equations must be non-linear.
- ○ At least one equation must be a non-horizontal line.
- ○ The *line* option found under *draw* in the calculator may only be used to make vertical lines.
- ○ The *circle* option should not be used.
- ○ You might need to limit your equations to achieve the picture you want.

To complete this project, students have to work in groups to answer guiding questions, such as the following:

- ○ Why do you need to program two equations to form a circle?
- ○ What makes a line move up and down?
- ○ Why is setting the window of the calculator so important?
- ○ The *TI-73* and *TI-83/84 Plus* graphing calculators contain random number generators that can allow the programmer to create a simulation model for the aforementioned "Bag of Beads" activity (McNamara, 2006, pp. 122–125). (The Bag of Beads activity is the 14th activity in Chapter 7: *Manipulatives.*

Below is a sample of a short *TI-82* program written by a student that will draw a smiley face:

```
: ClrDraw

: AxesOff

: −9 → Ymin

: 12 → Xmax

: −12 → Xmin

: 9 → Ymax
```

: ZSquare

: DrawF ($\sqrt{\ }(-x^2 + 25))-1$

: Draw F $(-\sqrt{\ }(-x^2 + 25)) + 1$

: DrawF $(.3x^2 -3/(x \leq 3) (-3 \leq x)$

: Line $(-1, 1, -1, 3)$

: Line $(1, 1, 1, 3)$

: Draw F (9)

: Draw F (-9)

: Vertical -13.6

: Vertical 13.6

(Connie Matchell, Geometry and Algebra II Teacher, Siloam Springs Public Schools, Siloam Springs, AR)

WHO: Grade 5–Grade 8
WHEN: During the lesson
FOCAL POINT(S):   Geometry

- Students can employ programs such as the Geometer's Sketchpad and explore a myriad of geometry concepts from basic to advanced. Students can also use the program to conceptually (or informally) prove theorems, such as the Pythagorean Theorem, by building right triangles and manipulating them to change the length of the sides, noting that the equation of the theorem remains constant.

REFLECTION AND APPLICATION

> ### How will I incorporate *technology* into mathematics instruction with my students?

Concept _____

Activity _____

_____

_____

Concept _____

Activity _____

_____

_____

Concept _____

Activity _____

_____

_____

Concept _____

Activity _____

_____

_____

Concept _____

Activity _____

_____

_____

Visualization and Guided Imagery

WHAT: DEFINING THE STRATEGY

My niece, Catherine, is a phenomenal fast pitch softball pitcher for her high school. She has appeared several times on the sports pages of the *Atlanta Journal–Constitution,* and her no-hitters continue to deserve coverage from the media and attention from college sports teams nationwide. The main secret of her success is her ability to visualize herself throwing strikes. "Cat" as we call her, throws strikes about 70% of the time or more and at times is simply unhittable. Her coach has taught her the strategy of visualization since he knew it would improve her performance.

Coaches everywhere tell me that it is a good thing to have their athletes visualize themselves hitting the home run or making the field goal prior to the game beginning. I understand that golfer Tiger Woods visualizes every ball going into every hole before he hits it and tennis player Roger Federer visualizes himself winning every game prior to taking the court. According to Stephen Covey (1996), everything is created twice, first in the mind and then in real life. I relate the same thing in rhyme:

> *If I can see it in the brain*
> *I may do it in the game.*

Students in Singapore have some of the highest math scores in the world. One strategy used with these students is to visualize a math problem before they attempt to solve it. When you have your students, like those in Singapore, use all of their senses (visual, auditory, tactile, and kinesthetic) to picture a problem, they stand a better chance of being able to solve it.

WHY: THEORETICAL FRAMEWORK

A picture in your mind creates a memory you can find (Sprenger, 2006b). Students who use the following four types of visualization acquire concepts more easily than those who rely only on auditory methods: (Posamentier & Jaye [2005, p. 74])

Type 1	Listen	Write	Picture
Type 2	Read	Write	Picture
Type 3	Read	Picture	
Type 4	Listen	Picture	

Since visualization is a major method to assist students in problem solving, a visualization technique guided by the teacher can help students fill the gap between concrete and abstract problem solving (Bender, 2005).

When the brain visualizes or forms mental images, the very same sections of the brain's visual cortex are activated as when the eyes actually process input from the real world, thereby enhancing retention and learning (Sousa, 2006).

When students intentionally construct a visual image of a math problem, they strengthen their understanding of the problem and increase their ability to think abstractly (Bender, 2005).

It is essential that a student visualize math problems because the visual cortex is involved in the majority of mathematical thinking (Sousa, 2001).

One of the four major underlying concepts that serve as a unifying idea in geometry is visualization, the ability to create visual images or use spatial reasoning when solving problems (Gavin, Belkin, Spinelli, & St. Marie, 2001).

A meta-analysis of 1,500 students representing nine separate studies showed that those who visualized or used more mental imagery while learning engaged in creativity during discussions, modeling, and assessments (LeBoutillier & Marks, 2003).

Students in the early years of schooling should be capable of developing visualization skills through firsthand experiences with various geometric objects (National Council of Teachers of Mathematics, 2000).

HOW: INSTRUCTIONAL ACTIVITIES

WHO: Prekindergarten–Grade 8
WHEN: During the lesson
FOCAL POINT(S):   All

• To provide practice in visualizing, as you read a word problem aloud, have students imagine each step of the problem. Have them see in their mind what is happening and then determine what operations are needed to solve the problem. Stop periodically and have students draw what they are visualizing.

WHO: Grade 1–Grade 8
WHEN: During the lesson
FOCAL POINT(S):   All

- As your students are reading and solving math problems independently, have them picture each step of the problem in their minds. Everything happens twice, once in the mind, and once in reality!

WHO: Grade 3–Grade 5
WHEN: During the lesson
FOCAL POINT(S):   Geometry

- Have students comprehend attributes and properties of two-dimensional figures by having them visualize what their three-dimensional counterparts would look like.

WHO: Grade 3–Grade 8
WHEN: During the lesson
FOCAL POINT(S):   Geometry

- Have students visualize what a given shape would look like rotated 180 degrees, flipped vertically, or turned 90 degrees. Have them describe or draw the resulting figure.

WHO: Grade 1–Grade 8
WHEN: After the lesson
FOCAL POINT(S):   All

- To alleviate anxiety prior to a math test, have students take deep breaths and visualize themselves successfully completing each item on the test. This activity, in addition to well-taught lessons incorporating the brain-compatible strategies, gives students the confidence they need to do well.

WHO: Grade 2–Grade 5
WHEN: During the lesson
FOCAL POINT(S):   Geometry

- Put the students in pairs. Give one student a shape of almost any two-dimensional design. Ask this student to describe the shape (in steps) to the other student who then must try to replicate the shape from the oral description. When the drawing is done, the students can compare the original shape to the drawn shape, discuss, get a new shape, and switch roles.

WHO: Grade 1–Grade 5
WHEN: During the lesson
FOCAL POINT(S):   Number and Operations

- Exercising the mind helps build visualization skills; so even engaging students in doing mental computation teaches them how to depend on the workings of their own mind to participate mathematically.

REFLECTION AND APPLICATION

> ## How will I incorporate *visualization* into mathematics instruction with my students?

Concept _____

Activity _____

_____

_____

Concept _____

Activity _____

_____

_____

Concept _____

Activity _____

_____

_____

Concept _____

Activity _____

_____

_____

Concept _____

Activity _____

_____

_____

STRATEGY 18

Visuals

WHAT: DEFINING THE STRATEGY

There are so many visual stimuli in the world that the visual cortex of students today is actually thicker than it was in my brain when I was their age. Look at all of the information that is coming into students' brains through the visual modality. They look at the computer screen. They play video games. They watch television. The visual modality is the strongest modality for a great many students in your classroom. This is the reason you need visuals; things that students can actually see.

Try reading a math word problem aloud to students and watch how many ask for a visual copy of it. Let's reflect on the Chinese proverb again.

> *Tell me, I forget.*
> *Show me, I remember.*
> *Involve me, I understand!*

While student engagement in the strategies of role play, movement, projects, or games is surely preferable for long-term retention of information, showing students what you are teaching them is still preferable to just telling them. Just be sure that the visuals connect to the auditory information being presented. Otherwise they become a distraction.

WHY: THEORETICAL FRAMEWORK

Visual learners take in the world through pictures and words and need to see the teacher solve a problem first to understand it (Sprenger, 2006b).

For English learners, visual tools in math offer visual ways of thinking about relationships and communicating information (Coggins et al., 2007).

> When visuals and the auditory information used to explain those visuals go together, they can be helpful. Otherwise pictures can interfere with a person's ability to listen to the words (Posamentier & Jaye, 2005).
>
> Visual learners take in the world through words and through pictures (Sprenger, 2006a).
>
> Even though rote learning plays some part, students in Singapore comprehend abstract concepts by using visual tools (Prystay, 2004).
>
> The visual cortex is involved in mathematical understanding since students need to see most math problems prior to solving them (Bender, 2005).
>
> When students are comparing and contrasting representations and solution methods, visual tools can assist in developing conceptual understanding (National Research Council, 2001).
>
> Models and pictures provide concrete referents to assist students in developing clearer mathematical understandings that can be shared with their peers (Hiebert et al., 1997).

HOW: INSTRUCTIONAL ACTIVITIES

WHO: Kindergarten–Grade 8
WHEN: During the lesson
FOCAL POINT(S):   All

•   When introducing a new math concept, work at least three problems on the board so that all students can see the steps involved in the solution. Most brains need a minimum of three examples before they begin to understand the procedure. As you solve each problem, model your thought processes aloud so that students can begin to understand the thinking involved.

WHO: Kindergarten–Grade 8
WHEN: During the lesson
FOCAL POINT(S):   All

•   Have students come to the chalkboard, dry erase board, overhead, or SMART Board and work problems that can serve as a visual for the remainder of the class. Have them explain the steps in solving the problem so that students have an auditory link to the visual problem.

WHO: Grade 1–Grade 5
WHEN: Before the lesson
FOCAL POINT(S):   Number and Operations

•   Make a sign similar to the one below to post on the wall or a bulletin board as a visual reminder to students of the cue words to look for in solving word problems. Remind students that this is just one of many strategies to use for solving problems and, like other strategies, may not be a good fit for every problem.

Operation	Terms Used
Addition	Altogether, add, how many, put together, in all
Subtraction	Take, took away, left, gave away
Multiplication	Problems that tell about one and then ask for total
Division	Problems that tell about many and then ask about one

WHO: Kindergarten–Grade 2
WHEN: During the lesson
FOCAL POINT(S):   Measurement

• Place cup, pint, quart, and gallon containers on a table for students to see. Demonstrate to students various measurements by pouring a liquid (water or Kool-Aid) from a gallon container into a smaller one. Ask them to estimate how many quarts it would take to hold the liquid in this gallon container. Show them that you would have to fill the quart container four times to use all of the liquid in the gallon. Ask students to estimate how many pints in the quart, how many cups in the pint, and so forth. Then pour the liquid as a visual reminder of each type of measurement.

WHO: Grade 3–Grade 5
WHEN: During the lesson
FOCAL POINT(S):   Number and Operations, Measurement

• Have students estimate the number of M&M's in a glass jar. Once students have had enough time to observe and place their estimates in a box, select the estimates, share them with the class, and discuss which ones are reasonable and unreasonable and why. Try the same experiment with various size jars.

(Bender [2005, p. 78])

WHO: Grade 1–Grade 8
WHEN: During the lesson
FOCAL POINT(S):   Data Analysis and Probability

• Draw bar and line-plot graphs on the board as visuals while you discuss data gathered in response to questions asked of students. For example, the following bar and line-plot graphs depict the number of siblings of the students in a second-grade class.

Brothers and Sisters We Have

Name	1	2	3	4	5
Aaron					
Betsy					
Christian					
David					
Elaine					
Franklin					
Grace					
Harriet					
Ivan					
Kenneth					
Lucy					
Matthew					
Nancy					
Oliver					
Pam					
Quincy					
Ricky					
Stan					
Wanda					
Vincent					
Yvette					

Siblings

Number of Students

	1	2	3	4	5
					X
					X
				X	X
			X	X	X
	X		X	X	X
	X	X	X	X	X
	X	X	X	X	X

WHO: Grade 1–Grade 3
WHEN: During the lesson
FOCAL POINT(S): Number and Operations

- Give each student a copy of the 100 chart (shown below) as a visual. Help them learn about number patterns by asking questions similar to the following:

  ○ *If you start at the number 23 and count by 10s, what number would you color next?*

  ○ *If you start at 48 and count by 10s, would 65 be one of the numbers you would include?*

(National Council of Teachers of Mathematics [2000, p. 93])

Number Table

1	2	3	4	5	6	7	8	9	10
11	12	13	14	15	16	17	18	19	20
21	22	23	24	25	26	27	28	29	30
31	32	33	34	35	36	37	38	39	40
41	42	43	44	45	46	47	48	49	50
51	52	53	54	55	56	57	58	59	60
61	62	63	64	65	66	67	68	69	70
71	72	73	74	75	76	77	78	79	80
81	82	83	84	85	86	87	88	89	90
91	92	93	94	95	96	97	98	99	100

WHO: Grade 3–Grade 8
WHEN: During the lesson
FOCAL POINT(S): Algebra

- Write the steps in an algebra problem on an overhead or dry erase board as you orally describe each step so that students can see it visually.

WHO: Grade 6–Grade 8
WHEN: During the lesson
FOCAL POINT(S):   Algebra

- Provide students with a written model solution (examples that you have worked out) to refer to when they are solving algebra problems on their own. These model solutions can serve as visuals to guide the thinking of your students and assist them when confusion exists.

WHO: Grade 6–Grade 8
WHEN: During the lesson
FOCAL POINT(S):   Geometry

- Discuss with students the fact that three-dimensional objects have parts that cannot be seen. Have them practice identifying unseen parts of figures in the visuals you provide by asking the following questions:
  - How many blocks in the figure are hidden from view?
  - How many total blocks are in the figure?
  - How many blocks would be hidden from view if we made the structure two blocks taller? Two blocks wider?

WHO: Grade 6–Grade 8
WHEN: During the lesson
FOCAL POINT(S):   Data Analysis and Probability

- Have students search print and visual media to find real-world examples of instances in which data has to be analyzed in order to reach specific conclusions. Have them bring in several examples to share with their cooperative group or with the entire class.

REFLECTION AND APPLICATION

How will I incorporate *visuals* into mathematics instruction with my students?

Concept _____

Activity _____

_____

_____

Concept _____

Activity _____

_____

_____

Concept _____

Activity _____

_____

_____

Concept _____

Activity _____

_____

_____

Concept _____

Activity _____

_____

_____

STRATEGY 19

Work Study and Apprenticeships

WHAT: DEFINING THE STRATEGY

My daughter, Jessica, finished college with a degree in German but had not found her passion. I asked her what career would ensure that she would look forward to going to work every day. She stated that what she really wanted was to one day own her own restaurant. "If that is the case," I said, "then, you must prepare for that day." So, she enrolled at the College of Culinary Arts at Johnson and Wales University and eventually completed another degree. Her course of study included a work-study program. She was placed at the Biltmore Estates in Asheville, North Carolina, and rotated for six months through the four restaurants on the grounds of that prestigious place. Jess learned more in six months while involved in that internship than she ever could have experienced sitting in a class.

Here is another example! A new seafood market opened in my neighborhood years ago. The market was owned by a family, and the middle-school age son was being taught by his father to ring up purchases. He was having a great deal of difficulty because the father was forcing him to estimate the change that the customers would receive prior to putting the numbers in the cash register. When I returned to the market in a few weeks, the son had gotten so good at figuring out the exact change to return to the customer that he hardly even needed the register at all. His brain had practiced a skill that was useful in the real world and had actually applied that skill in a real-world setting.

You see, Aristotle told us many years ago that *one learns to do by doing*. When students are shadowing a professional whose job involves the real-world application of math or when they are placed in a position which necessitates the practical use of mathematical concepts in the everyday world, learning is increased and retained.

WHY: THEORETICAL FRAMEWORK

Students should learn that mathematics is an ever-changing subject and that what we learn in school is related to the discoveries of real mathematicians and to everyday life (Rothstein, Rothstein, & Lauber, 2006).

Students are motivated when teachers show them the connections between mastery in mathematics and their success in other subjects, as well as math's relevance in their daily lives (Posamentier & Jaye, 2005).

Mathematics isn't a scary and abstract mystery when everyday life applications are used to teach it (Posamentier & Jaye, 2005).

The schoolwork of adolescents must take them into the "dynamic life of their environments" (Brooks, 2002, p. 72).

Educated adults often have difficulty finding a job or meeting job expectations since large gaps can exist between the performance needed to be successful in a business setting and that required for school success (Sternberg & Grigorenko, 2000).

The strongest neural networks are created when students are actually engaged in real-life experiences and not from tasks that are not authentic (Westwater & Wolfe, 2000).

Learning should be organized around cognitive-apprenticeship principles that stress subject-specific content and the skills required to function within the content (Berryman & Bailey, 1992).

HOW: INSTRUCTIONAL ACTIVITIES

WHO: Kindergarten–Grade 2
WHEN: During the lesson
FOCAL POINT(S):   Number and Operations

* **Best buy!**—Have students bring in empty boxes, cans, or bottles to include in the classroom grocery store. Before opening the store, review age-appropriate money concepts: the worth of a coin; counting by 1s, 5s, 10s, 25s; making change; and so forth. Generate a discussion on how items are arranged in the grocery store and determine the price for each item. Tag each item with a price and establish the rules for making purchases. Open the classroom store for operation. Students can take turns being the cashier and/or restocking the store items. This can be either a center or a whole class activity.

(Tomiko T. Smalls, Grade 2, Mossy Creek Elementary School, North Augusta, SC.)

WHO: Grade 3–Grade 8
WHEN: After the lesson
FOCAL POINT(S):   All

* Locate professionals who use mathematics in their work world. Bring one or two of these people in to talk to the class about the real-world application of mathematical concepts in their profession.

WHO: Grade 6–Grade 8
WHEN: After the lesson
FOCAL POINT(S):   All

  • Partner with local businesses who can make it possible for students to engage in internships, apprenticeships, and work-study projects (either during the school year or during the summer months) so that students can experience firsthand the knowledge and skills essential for the workplace. Allow them to spend time with professionals who use mathematics in their daily occupations.

WHO: Grade 3–Grade 8
WHEN: During the lesson
FOCAL POINT(S):   All

  • Engage students in a service learning project where they provide a service for their school or community while simultaneously mastering a curricula focal point in math. For example, have them beautify the school grounds by planning and implementing a butterfly garden. Have them research the necessary components for the garden, perform the essential measurements of the grounds, and calculate what should be planted where, while journaling the entire experience. Service learning is one of the best vehicles for combining interdisciplinary instruction with real-world skills and strategies (Tate, 2003).

WHO: Grade 6–Grade 8
WHEN: After the lesson
FOCAL POINT(S):   Number and Operations

  • Have students take turns working as apprentices in the school store to develop the knowledge and skills necessary to become entrepreneurs (Tate, 2003).

WHO: Grade 6–Grade 8
WHEN: During the lesson
FOCAL POINT(S):   Number and Operations

  • Tell students they have been placed in a time machine and are now 18 years old. They have been given $50,000 and will need to prepare a budget to meet all of their basic needs for one year. This project will require a great deal of research into the cost of living arrangements, average monthly electric, gas, and water bills, which can then be rounded. Students then should create a chart on the computer that shows the money being subtracted from the $50,000 down to zero.

(Darla R. Quinn, Grade 5, Janet Johnstone Elementary School, Calgary, Alberta, Canada)

WHO: Grade 6–Grade 8
WHEN: After the lesson
FOCAL POINT(S):   Number and Operations

- Under the direction of a teacher, have students plan and operate a school bank in which fellow students can deposit money. Use this real-life experience to enable students to master math concepts such as bank deposits and withdrawals, interest rates, percentages, and so forth. Have students take turns serving as apprentices in the banking business (Tate, 2003).

WHO: Grade 5–Grade 8
WHEN: Before the lesson
FOCAL POINT(S):   All

- *The Eddie Files,* published by PBS, documents a student as he learns how professionals use mathematics in the workplace. This is a resource that might be helpful to inspire students to be open to the ways people use mathematics to be productive.

REFLECTION AND APPLICATION

> **How will I incorporate *work study* and *apprenticeships* into mathematics instruction with my students?**

Concept _____

Activity _____

_____

_____

Concept _____

Activity _____

_____

_____

Concept _____

Activity _____

_____

_____

Concept _____

Activity _____

_____

_____

Concept _____

Activity _____

_____

_____

Writing and Journals

WHAT: DEFINING THE STRATEGY

Writing something down appears to help us remember. For example, I have made lists of items to get when shopping at the mall, never even looked at my list, but managed to purchase the items written down. However, I have been in middle and high school classrooms where teachers were asking students to take copious notes but were continuing to talk to students at the same time. The brain appears to be able to give conscious attention to only one thing at a time. So students are often missing part of what the teacher is saying as they are writing notes or they are missing notes while attempting to listen to the teacher. By the way, if the brain can only pay conscious attention to one thing at a time, what does that say for talking on your cell phone while driving? Even more ludicrous is the teenager who text messages while driving. Well, I digress! Let's get back to the strategy of writing.

Writing is crucial to understanding and memory and is cross curricular. It should be integrated into every single discipline, including math. Having students write out the steps when solving a computational or word problem or write about their thinking in a math journal goes a long way toward increasing math achievement.

WHY: THEORETICAL FRAMEWORK

The following benefits should be considered when engaging students in writing activities in math:

- Thoughts are permanently recorded for later reflection
- Knowledge is internalized and can be applied in areas of interest
- Confidence and a sense of self-understanding is gained
- Problem solving and critical-thinking skills are developed
- Students can determine what is intellectually relevant to them (Sousa [2007])

When writing and mathematics instruction are integrated at every grade level, students learn to express their concepts and ideas while thinking mathematically, and student achievement levels increase (Rothstein, Rothstein, & Lauber, 2006).

When the kinesthetic activity of writing is used to communicate math concepts, more neurons are engaged and students are made to organize their thoughts (Sousa, 2007).

When students were given written model solutions (examples that had been worked out) to refer to when solving practice problems, they made fewer errors than a comparable group who solved a greater number of practice problems without the written model solutions (Posamentier & Jaye, 2005).

Prekindergarten through second grade students should be encouraged to use paper and pencil to record what they are thinking when solving computational problems (National Council of Teachers of Mathematics, 2000).

Having students write down what is observed, presented, or thought about helps the brain organize and make sense of extremely complicated and multifaceted bits of information (Jensen, 2000).

Students in kindergarten through Grade 12 should be able to do the following:

- Use communication to organize and consolidate their thinking
- Communicate to their classmates, teachers, and others their coherent mathematical thinking
- Evaluate the mathematical thinking of other people
- Use precise mathematical language to express mathematical ideas (National Council of Teachers of Mathematics [2000])

Students should be encouraged to talk and write about their ideas, to comprehend the basic concepts being taught, and to put those concepts into their own words (Kohn, 1999).

HOW: INSTRUCTIONAL ACTIVITIES

WHO: Grade 1–Grade 8
WHEN: During the lesson
FOCAL POINT(S):   Number and Operations

- Have students write in narrative form the thinking involved in solving computational problems. Since the brain tends to remember what it writes down, this activity should go a long way in ensuring that students remember the thinking behind the solutions they derive.

WHO: Grade 1–Grade 8
WHEN: During the lesson
FOCAL POINT(S):   All

- Have students write down the steps needed to solve word problems. Not only will the written steps assist the student in remembering the sequence of the solution but will also provide insight into the thinking of the student during problem solving.

WHO: Grade 1–Grade 8
WHEN: After the lesson
FOCAL POINT(S):   All

- Have students discuss a mathematical concept or talk about their ideas with a partner or group prior to writing about those ideas. Students could also work with a partner to make observations regarding the data reflected in a particular graph. Have them write one or two sentences that reflect the conclusions gleaned.

WHO: Grade 1–Grade 8
WHEN: During the lesson
FOCAL POINT(S):   All

- Incorporate *Quick Writes* (a technique for having students stop and quickly jot down an answer to a question) throughout a math lesson. Stop periodically during the lesson and have students write a concept just taught. Writing even for a minute will help to reinforce the content. For example, stop your lesson and tell students the following: *Write the three types of triangles that we are studying. Write five examples of prime numbers.*

WHO: Grade 1–Grade 8
WHEN: During the lesson
FOCAL POINT(S):   All

- In an effort to improve the quality of the journal writing of students, have students brainstorm a *Math Alphabet Book* where they include mathematical vocabulary chunked according to the letters of the alphabet, yet pertinent to a unit of study. For example, during a unit of geometry, a *Geometry Alphabet Book* could look like the following: *Acute, Base, Circumference, Diameter, Equilateral, Figure,* and so forth. Post these words as a visual or have students include them in their math notebooks for ready reference.

WHO: Grade 1–Grade 8
WHEN: During the lesson
FOCAL POINT(S):   All

- Have students create original word problems incorporating several vocabulary words provided by the teacher. The vocabulary words can be selected from the *Alphabet Books* described above and should relate to the focal points previously taught. Allow students who have difficulty working alone to work in pairs or cooperative groups to create their problems. Students can exchange papers and solve one another's problems and develop a word wall for the vocabulary.

WHO: Grade 1–Grade 8
WHEN: After the lesson
FOCAL POINT(S):   All

- Following a unit of study in algebra, or any other focal point, have students record their thoughts regarding the unit in their personal journals. The following open-ended question starters may serve to spark the thinking of students:
    - State at least three major concepts you learned in this unit.
    - What was your favorite activity in which the class participated?
    - What was your least favorite activity in which the class participated?
    - How can you apply what you have learned to your personal life or to a future career choice?
    - What things would you change if this unit were taught again?

WHO: Grade 3–Grade 8
WHEN: During the lesson
FOCAL POINT(S):   All

- Have students research and write reports regarding the role math plays in everyday life or the life of a mathematician and that person's contribution to society.

REFLECTION AND APPLICATION

> **How will I incorporate *writing* and *journals* into mathematics instruction with my students?**

Concept _____

Activity _____

_____

_____

Concept _____

Activity _____

_____

_____

Concept _____

Activity _____

_____

_____

Concept _____

Activity _____

_____

_____

Concept _____

Activity _____

_____

_____

Resource A

Have You Read Any Math Lately?

Adler, D. A. (1999). *How tall, how short, how far away?* (Grades 3–5). New York: Holiday House. ISBN 0-8234-1632-1.
Skills: Standard & Metric Measurement.

Axelrod, A. (1998). *Pigs on a blanket: Fun with math and time.* (Ages 4–9). New York: Aladdin. ISBN 0-689-82252-9.
Skills: Elapsed Time.

Axelrod, A. (1999). *Pigs go to market: Fun with math and shopping.* (Ages 4–9). New York: Aladdin. ISBN 0-689-82553-6.
Skills: Weights, Measures, & Multiplication.

Axelrod, A. (1999). *Pigs on the move: Fun with math and travel.* (Ages 4–9). New York: Simon & Schuster. ISBN 0-689-81070-9.
Skills: Time & Distance.

Axelrod, A. (1999). *Pigs in the pantry: Fun with math and cooking.* (Ages 4–9). New York: Aladdin. ISBN 0-689-82555-2.
Skills: Liquid & Solid Measures.

Axelrod, A. (2000). *Pigs on the ball: Fun with math and sports.* (Ages 4–9). New York: Aladdin. ISBN 0-689-83537-X.
Skills: Geometry.

Axelrod, A. (2003). *Pigs at odds: Fun with math and games.* (Ages 4–9). New York: Aladdin. ISBN 0-689-86144-3.
Skills: Probability.

Birch, D. (1993). *The king's chessboard.* (Grades 4 & up). New York: Penguin. ISBN 0-14-054880-7.
Skills: Exponential Growth.

Ellis, J. (2004). *What's your angle, Pythagoras? A math adventure.* (Grades 4 & up). Watertown, MA: Charlesbridge. ISBN 1-57091-150-9.
Skills: Pythagorean Theorem.

Fadiman, C. (1997). *Fantasia mathematica*. (Adult reading; easily adapted for storytelling to students). New York: Springer. ISBN 0-387-94931-3.
Skills: General Mathematics & Science.

Fadiman, C. (1997). *The mathematical magpie*. (Adult reading; easily adapted for storytelling for students). New York: Springer. ISBN 0-387-94950-X.
Skills: General Mathematics & Science.

Fox, M. (2001). *Reading magic*. New York: Harcourt. ISBN 0-15-601155-7. (This is an excellent teacher resource that examines the importance of reading aloud to all children.)

Glass, J. (1998). *The fly on the ceiling: A math myth*. (Grades 2–3). New York: Random House. ISBN 0-679-88607-9.
Skills: Coordinate Graphing.

Greenberg, D. (1999). *Mega-funny math poems and problems*. (Grades 3–6). New York: Scholastic. ISBN 0-590-18735-X.
Skills: Multiplication, Division, Fractions, Measurement.

Holtzman, C. (1997). *No fair!* (Grades K–2). New York: Scholastic. ISBN 0-590-92230-0.
Skills: Probability.

Hopkins, L. B. (1997). *Marvelous math: A book of poems*. (Ages 5 & up). New York: Aladdin. ISBN 0-689-84442-5.
Skills: Assorted.

Leedy, L. (1994). *Fraction action*. (Grades 2–4). New York: Holiday House. ISBN 0-8234-1244-X.
Skills: Introduction to Fractions.

Ling, B. (1997). *The fattest, tallest, biggest snowman ever*. (Grades 1–2). New York: Scholastic. ISBN 0-590-97284-7.
Skills: Measurement.

Maccarone, G. (1997). *Three pigs, one wolf, and seven magic shapes*. (Grades 1–2). New York: Scholastic. ISBN 0-590-30857-2.
Skills: Geometry.

McGrath, B. (1998). *More M&M's math*. (Grades 3–5). Watertown, MA: Charlesbridge. ISBN 0-88106-994-9.
Skills: Classifying, Graphing, Addition, Subtraction, Multiplication, Ordinal Numbers.

McGrath, B. (2000). *Skittles riddles math*. (Grades 4 & up). Watertown, MA: Charlesbridge. 2000. ISBN 1-57091-413-3.
Skills: Comparing Numbers, Computation, Positive/Negative Numbers, Fractions.

Moscovich, I. (2001). *1000 play thinks*. (All ages). New York: Workman. ISBN 0-7611-3158-2.
Skills: Geometry, Topology, Networks, Logical Thinking Skills.

Moscovich, I. (2001). *Network games*. (Ages 8–12). New York: Workman. ISBN 0-761-12019-X.
Skills: Networks, Critical & Logical Thinking Skills.

Moscovich, I. (2001). *Pattern games*. (Ages 8–12). New York: Workman. ISBN 0-761-12020-3.
Skills: Patterns, Critical & Logical Thinking Skills.

Moscovich, I. (2001). *Probability games*. (Ages 8–12). New York: Workman. ISBN 0-761-12017-3.
Skills: Probability, Critical & Logical Thinking Skills.

Moscovich, I. (2003). *The awesome 3-D puzzle challenge*. (All ages). New York: Sterling. ISBN 1-402-70709-6.
Skills: Critical & Logical Thinking Skills.

Murphy, S. J. (1996). *Too many kangaroo things to do*. (Ages 7 & up). New York: HarperCollins. ISBN 0-06-446712-0.
Skills: Multiplying.

Murphy, S. J. (1997). *The best vacation ever*. (Ages 6 & up). New York: HarperCollins. ISBN 0-06-446706-6.
Skills: Collecting Data.

Murphy, S. J. (1997). *Betcha*. (Grades 7 & up). New York: HarperCollins. ISBN 0-06-446707-4.
Skills: Estimating.

Murphy, S. J. (2000). *Dave's down-to-earth rock shop*. (Ages 7 & up). New York: HarperCollins. ISBN 0-06-446729-5.
Skills: Classifying.

Murphy, S. J. (2001). *Probably pistachio*. (Ages 6 & up). New York: HarperCollins. ISBN 0-06-446734-1.
Skills: Probability.

Murphy, S. J. (2002). *Safari park*. (Ages 7 & up). New York: HarperCollins. ISBN 0-06-446245-5.
Skills: Finding Unknowns.

Murphy, S. J. (2003). *The grizzly gazette*. (Ages 7 & up). New York: HarperCollins. ISBN 0-06-000026-0.
Skills: Percentage.

Murphy, S. J. (2003). *Less than zero*. (Ages 7 & up). New York: HarperCollins. ISBN 0-06-000126-7.
Skills: Negative Numbers.

Murphy, S. J. (2004). *Treasure map*. (Ages 3 & up). New York: HarperCollins. ISBN 0-06-446738-4.
Skills: Mapping, Interpreting Symbols, Understanding Direction & Scale.

Myller, R. (1991). *How big is a foot?* (Grades 1–3). Yearling. ISBN 0-440-40495-9.
Skills: Measurement.

Neuschwander, C. (1997). *Sir cumference and the first round table*. (Grades 3–6). Watertown, MA: Charlesbridge. ISBN 1-57091-152-5.
Skills: Geometry.

Neuschwander, C. (1998). *Amanda Bean's amazing dream: A mathematical story*. (Grades 2–4). New York: Scholastic. ISBN 0-590-30012-1.
Skills: Multiplication.

Neuschwander, C. (1999). *Sir cumference and the dragon of Pi*. (Grades 3–6). Watertown, MA: Charlesbridge. ISBN 1-57091-164-9.
Skills: Geometry.

Neuschwander, C. (2001). *Sir cumference and the great knight of Angleland*. (Grades 3–6). Watertown, MA: Charlesbridge. ISBN 1-57091-169-X.
Skills: Geometry.

Neuschwander, C. (2003). *Sir cumference and the sword in the cone*. (Grades 3–6). Watertown, MA: Charlesbridge. ISBN 1-57091-601-2.
Skills: Geometry.

O'Donnell, K. (2004). *Space circles: Learning about radius and diameter*. (Grades 4 & up). New York: Rosen. ISBN 0-8239-8878-3.
Skills: Radius and Diameter.

Pallotta, J. (2001). *Twizzlers percentage book*. (Ages 6–9). New York: Scholastic. ISBN 0-439-15430-8.
Skills: Fractions, Decimals, Percents.

Pallotta, J., & Bolster, R. (1999). *The Hershey's milk chocolate fractions book*. (Grades 3 & up). New York: Scholastic. ISBN 0-439-13519-2.
Skills: Fractions.

Pappas, T. (1993). *Fractals, googols, and other mathematical tales*. (Grades 4 & up). San Carlos, CA: Wide World Publishing. ISBN 0-933174-89-6.
Skills: Geometry, Topology & Networks, Patterns, Pi, & Fractals.

Pappas, T. (1997). *The adventures of Penrose, the mathematical cat*. (Grades 4 & up). San Carlos, CA: Wide World Publishing. ISBN 1-884550-14-2.
Skills: Geometry, Abacus, Tangrams, Probability, Magic Squares, Tessellations, Multiplication.

Pinczes, E. J. (1997). *A remainder of one.* (Ages 7 & up). New York: Scholastic. ISBN 0-590-12705-5.
Skills: Division, Fact Families.

Pluckrose, H. (1995). *Math counts: Pattern.* (Ages 4 & up). New York: Children's Press. ISBN 0-516-45455-2.
Skills: Recognizing Patterns.

Rocklin, J. (1998). *The case of the backyard treasure.* (Grades 2–3). New York: Scholastic. ISBN 0-590-30872-6.
Skills: Problem Solving, Measurement, Geometry.

Rocklin, J. (1999). *The case of the $hrunken allowance.* (Grades 2–3). New York: Scholastic. ISBN 0-590-12006-9.
Skills: Money, Time, Estimation, Measurement, Logical Reasoning.

Schwartz, D. M. (1994). *If you made a million.* (Grades 3 & up). New York: HarperCollins. ISBN 0-688-13634-6.
Skills: Earning Money, Investing Money, Accruing Dividends & Interest.

Schwartz, D. M. (1998). *G is for googol: A math alphabet book.* (Grades 4 & up). Berkeley, CA: Tricycle Press. ISBN 1-883672-58-9.
Skills: Geometry, Patterns, Probability, Graphing.

Schwartz, D. M. (1999). *The magic of a million.* (Grades 2–5). New York: Scholastic. ISBN 0-590-70133-9.
Skills: Place Value, Number Sense.

Schwartz, D. M. (2001). *On beyond a million: An amazing math journey.* (Ages 6–10). Oklahoma: Dragonfly Books. ISBN 0-440-41177-7.
Skills: Exponents.

Scieszka, J. (1995). *Math curse.* (Ages 6–99). New York: Viking Press. ISBN 0-670-86194-4.
Skills: Time, Patterns, Money, Computation, Fractions.

Tompart, A. (1997). *Grandfather Tang's story.* (Grades 3 & up). Oklahoma: Dragonfly Books. ISBN 0-517-88558-1.
Skills: Geometry.

Wells, R. E. (1993). *Is a blue whale the biggest thing there is?* (Grades 3 & up). Morton Grove, IL: Albert Whitman & Co. ISBN 0-8075-36563.
Skills: Place Value.

Wells, R. E. (2000). *Can you count to a googol?* (Grades 3 & up). Morton Grove, IL: Albert Whitman & Co. ISBN 0-8075-1061-0.
Skills: Place Value, Multiples of 10.

(Compiled by Karen S. Arnett, Elementary Gifted Education Specialist, Chesapeake, Virginia and based on Whitin, D. & Wilde, S. [1992]. *Read any good math lately?* Heinemann, a division of Reed Elsevier, Inc.: Portsmouth, NH.)

Resource B

Brain-Compatible Lesson Design

So now what? In other words, how do you incorporate the 20 brain-compatible strategies into your daily mathematics lesson plans? A sample lesson plan is displayed in Figure B.1 on page 154. Each of the major sections of this plan is described in detail in the paragraphs that follow.

This completed lesson plan is merely one example of how the strategies can be integrated into a lesson that not only increases student achievement and long-term retention but also enables students to have fun while learning mathematics. An adaptation of this plan is currently being used nationwide as school systems revise curriculum and ensure that lessons are planned which maximize student achievement and help students meet content standards.

Section 1—Lesson objective: What will you be teaching?

Obviously, when teachers are planning math lessons, the first question they should ask themselves should be: *What will I be teaching?* This question is answered by examining curriculum that addresses local, state, and national standards; it all should reflect the focal points addressed in this text. Gone should be the days when math teachers have students open the textbook to page 1 at the beginning of the school year and conclude math instruction with the last page of the book prior to summer vacation.

A better way of teaching calls for a paradigm shift on the part of many professionals who look at math as isolated skills to be mastered. Recent research shows us that *less is more*. In fact, when researchers compared mathematics curriculum in the United States with countries that are more successful on international assessments, they found that the U.S. curriculum includes too many topics lacking the focus and intensity essential for mastery (Solomon, 2006). Teachers instead should be identifying the big ideas, patterns, or focal points in mathematics instruction and teaching those big ideas over and over in a variety of different contexts (Bender, 2005). For example, students in Singapore have some of the highest mathematics test scores of any students in the world. Yet mathematics textbooks

BRAIN-COMPATIBLE *Mathematics* LESSON PLAN

Lesson Objective(s): *What will you be teaching?*

Assessment (Traditional/Authentic): *How will you know students have learned the content?*

_____

Ways to Gain/Maintain Attention (Primacy): *How will you gain and maintain students' attention? Consider need, novelty, meaning, or emotion.*

_____

Content Chunks: *How will you divide and teach the content to engage students' brains?*

Lesson Segment 1:

Activities:

Lesson Segment 2:

Activities:

Lesson Segment 3:

Activities:

_____

Brain-Compatible Strategies: *Which will you use to deliver content?*

☐ Brainstorming/Discussion	☐ Drawing/Artwork	☐ Field Trips
☐ Games	☐ Graphic Organizers/Semantic Maps/Word Webs	☐ Humor
☐ Manipulatives/Experiments/Labs/Models	☐ Metaphor/Analogy/Simile	☐ Mnemonic Devices
☐ Movement	☐ Music/Rhythm/Rhyme/Rap	☐ Project/Problem-based Instruction
☐ Reciprocal Teaching/Cooperative Learning	☐ Role Play/Drama/Pantomime/Charades	☐ Storytelling
☐ Technology	☐ Visualization/Guided Imagery	☐ Visuals
☐ Work Study/Apprenticeships	☐ Writing/Journals	

Figure B.1

in Singapore are one-third the size of comparable grade-level textbooks in the United States (Prystay, 2004).

Teaching math in relevant chunks with the integration of specific activities is preferable. Math instruction taught in terms of the major grade-level focal points helps to ensure that students not only see the relationships between and among various concepts from one year to the next but also experience math in terms of what is happening in the real world.

Let's look again at Scenario II in the Introduction section of this text. If you remember, Mr. Rutledge expects the students in his class to know how to determine the mean, mode, and median of a group of numbers. That is his lesson objective and answers the question: *What will you be teaching?* Why don't we use this lesson and carry it through the brain-compatible lesson plan questions that follow?

Section 2—Assessment: How will you know students have learned the content?

When planning a math lesson, if you wait until the completion of the plan to decide how to assess your students, you have waited too late. I can still visualize myself as a student in school. I remember being stressed on math test day because assessment meant trying to guess what my teachers were going to put on their tests. If I guessed correctly, I would manage to make an "A." However, if I guessed incorrectly, even if I studied feverishly, my grade wasn't so good. Current research tells us to *begin with the end in mind.* Tell students what you expect: What they should know and be able to do at the culmination of a lesson or unit of study. In this way, assessment may challenge the brain, but not to the point of being a high stressor. Consider this analogy: How can a pilot file a flight plan without knowing the destination?

In our example, Mr. Rutledge told the students in his class that by the culmination of the lesson, they should be able to determine the mean, median, and mode of any given group of numbers.

Section 3—Ways to gain and maintain attention: How will you gain and maintain students' attention? (Consider need, novelty, meaning, and emotion.)

There are so many stimuli in the world that your brain becomes very particular about what gets its attention. When you are teaching, you are vying for a spot in the brains of your students. But your lesson may be competing with the conversation of a peer, a noise in the hall, a colorful leaf on a tree outside, or reflections of an argument that the student had with a family member before coming to school. Students can even be staring you in the face and not paying a bit of attention to what you are teaching.

There are four major ways to gain your students' attention. Mr. Rutledge has the benefit of using all four ways:

Need

The first way to grab the attention of your students is through *need*. If students see the purpose of what you are teaching, they will see the need to learn it. For example, I did not see the need to memorize the cell phone numbers of my three children since they were all programmed into my cell phone. However one day last year, my cell phone died. I was traveling and needed to talk to my oldest daughter. I couldn't get in touch with my husband and had not written down my children's numbers. All of a sudden, I saw a purpose in memorizing their numbers and since then, I have.

When students see the purpose of your math lesson, they will see the need to learn its content. Simply telling them they will need the information for a subsequent test may not be enough motivation for many students. However, showing them why the need to know how to calculate simple interest will come in handy when they are ready to transact a loan for a car may be just enough to capture their attention.

In our sample lesson, Mr. Rutledge gave students a purpose. They would need to know how to determine the mean, median, and mode of a set of song ratings to determine which student's song was best liked by the entire class.

Novelty

Another way to get students' attention is by teaching your content in a *novel* and interesting way. When the brain becomes accustomed to the same scenery over and over, it simply ignores some things. For example, I am on airplanes so much of the time that I simply pay little attention when the flight attendants give instructions. I hear them so often that I could actually mouth the instructions with the flight attendant. However one night I flew Southwest Airlines where the flight attendants were singing the instructions to the tune of *The Beverly Hillbillies.* Did I pay attention? You bet I did!

While you certainly want consistency in your class rituals and procedures, you will want to vary your lesson delivery. When you change your location in the room, your voice inflection, or the strategies you use to deliver your lesson, you are being novel, and you stand a better chance of gaining and maintaining your students' attention.

The 20 strategies provide you with many ways to be novel. Think of all the novel stories that you and your students can tell, the variety of songs you can play, the projects in which you can engage your students, and all the different movements you can use to put information into procedural memory. The options are endless!

In our sample lesson, Mr. Rutledge uses the activity of determining the favorite song of the class to teach this objective in a new and different, yet meaningful way. The lesson is truly novel!

Meaning

Since the brain's purpose is survival in the real world, when you connect your content to real life, you are making it meaningful. When you are not, students will raise their hands and ask the question: *Why do we have to learn this?* As I teach, I take every opportunity to use real-life examples to illustrate my points. I read lists of comedians who lived to over 80 years of age to illustrate the correlation between laughter and longevity. I show pictures of real people who look five to 10 years younger than their age to show what can happen when people have a positive attitude and stay active as they age.

Math content is no different. Take a concept you are teaching and formulate a problem that students would have to solve in their real world. You can even work one or two of your students' names into the problem itself. All of a sudden, the content becomes more meaningful because students can see how the concept relates to real life.

It is very meaningful for students to listen to their own music in our sample lesson. What stronger connection to real life can there be?

Emotion

Of all the ways to gain and maintain students' attention, emotion may be the most powerful! Emotion places memories in reflexive memory, one of the strongest memory systems in the brain, and helps to ensure retention. In fact, you will not soon forget anything that happened to you in your personal life or in the world at large that was emotional. I bet you can remember where you were on January 28, 1986, when you were informed the Challenger had exploded.

Teachers who are emotional about their content are passionate and enthusiastic! If you ever had a math teacher who loved math, no doubt their emotion was so contagious that you began to love math too; even if you didn't love it at the beginning of the school year.

In our example, Mr. Rutledge's lesson was an actual lesson taught in a middle school where I was observing. Students were so excited about this lesson that they were running to math class. In many schools I visit, students are running away from math class. So I followed the students and found out about the song contest. Their emotion was obvious!

Need, novelty, meaning, and *emotion* are four ways to gain the brain's attention. You do not need to have all four working for you in a lesson. Even one, used appropriately, will work as you compete with the multitude of stimuli surrounding your students during your lesson presentation.

Section 4—Content chunks: How will you divide and teach the content to engage students' brains?

Many years ago Madeline Hunter, a guru in the field of education, asked this question: *How do you eat an elephant?* The answer of course was: *One bite at a time.* Because the brain can only hold approximately seven bits of information (plus or minus two) simultaneously, the way to get students to remember more content is to connect or chunk it together. This is why the social security number is in chunks: It has nine digits and therefore needs to be connected together into three chunks. A 10-digit telephone number (with area code) is also divided into three chunks to make it easier for you to remember. You see, the brain remembers a chunk as if it were one piece of information, rather than separate numbers.

In our sample lesson, the objective was divided into the following three smaller chunks: how to determine the *mean*, the *mode*, and the *median*.

I have added another question to Madeline Hunter's question: *How do you digest an elephant?* The answer of course is that you have to *chew it up*. Activity enables the brain to *chew up* information. Chew is a metaphor for the fact that the activity enables the brain to process what it is learning. A classroom where there is little opportunity for students to process what they are learning is a math classroom where students are not performing at optimal levels and may not be comprehending or retaining as much information as they could.

Mr. Rutledge used several different activities in teaching each of the three chunks. The students listened to the five musical selections. To teach the *mode*, he had students work with a partner to figure out the most frequently occurring score. To teach the *mean*, he asked questions of volunteers and non-volunteers as he pulled some of the sample data and calculated the average. To teach the *median*, students watched on the SMART Board as he lined up the scores for one song to determine the middle score. Students then stood up and kept an appointment with a fellow student to take another song and follow the same procedure. They had to be finished determining the median score before the music ended. Students then wrote in their math journals the procedures for determining all three: the mean, mode, and median.

Section 5—Brain-compatible strategies: Which will you use to deliver content?

By the time a teacher completes a lesson plan, the activities included in the lesson should reflect the 20 brain-compatible strategies outlined in this book. In fact, on the bottom of the sample lesson plan form, all 20 strategies are listed so that teachers have a ready reference for their use.

In every lesson I teach, regardless of which grade level or content area, I attempt to incorporate at least four of the strategies, one from each of the four modalities: visual, auditory, kinesthetic, and tactile. (Refer to Figure I in the Introduction for a correlation of the strategies to the learning modalities.) In this way, regardless of student preferences, there is an activity in the lesson for every student, and instruction can be differentiated based on students' learning needs.

In our sample lesson, the strategies that were used to teach the objective include the following: *music* (students listened to their own music), *visuals* (students worked sample problems on the SMART Board), *reciprocal teaching* (students worked with a close partner to determine the mode), *movement* (students got up and kept an appointment with another classmate who sits at a distance), *problem-based instruction* (students worked with real-life data to solve a meaningful problem), lastly, *writing* (students wrote in their math journals describing the process for determining each concept taught). By the time this lesson ended, students had used at least six of the 20 brain-compatible strategies.

If you come to the end of a lesson plan and have not utilized any of the brain-compatible strategies, go back and plan it again. It is not brain-compatible! We must begin to teach smarter, not harder. Teaching smarter means the following:

1. Teaching the major focal points as chunks of information

2. Letting students know what is expected of them by the end of the lesson

3. Utilizing ways to gain and maintain the attention of students throughout the lesson because students tend to remember what happens first in your lesson

4. Deciding how many different chunks are needed to get students through the content

5. Incorporating brain-compatible strategies into the activities designed to teach each chunk of information

Bibliography

Allen, R. H. (2001). *Impact teaching: Ideas and strategies for teachers to maximize student learning.* Boston: Allyn & Bacon.

Bender, W. N. (2005). *Differentiating math instruction: Strategies that work for K–8 classrooms!* Thousand Oaks, CA: Corwin Press.

Berryman, S. E., & Bailey, T. R. (1992). *The double helix of education and the economy.* New York: Institute on Education and the Economy, Columbia University Teachers College.

Brooks, J. (2002). *Schooling for life: Reclaiming the essence of learning.* Alexandria, VA: Association for Supervision and Curriculum Development.

Brooks, J., & Brooks, M. (1993). *In search of understanding: The case for constructivist classrooms.* Alexandria, VA: Association for Supervision and Curriculum Development.

Bulla, D. (1996). *Think math! Interactive loops for groups.* Chicago: Zephyr Press.

Burden, P. R. (2000). *Powerful classroom management strategies: Motivating students to learn.* Thousand Oaks, CA: Corwin Press.

Burke, M., Erickson, D., Lott, J. W., & Obert, M. (2001). *Navigating through algebra in Grades 9–12.* Reston, VA: National Council of Teachers of Mathematics.

Burns, M. (1975). *The I hate mathematics! book.* New York: Little, Brown.

Caine, R. N., & Caine, G. (1994). *Making connections: Teaching and the human brain.* Menlo Park, CA: Addison-Wesley.

Catterall, J., Chapleau, R., & Iwanga, J. (1999, Fall). *Involvement in the arts and human development: Extending an analysis of general associations and introducing the special cases of intense involvement in music and in theater arts* (Monograph Series No. 11). Washington, DC: Americans for the Arts.

Checkley, K. (1999). *Math in the early grades: Laying a foundation for later learning.* Alexandria, VA: Association for Supervision and Curriculum Development.

Coggins, D., Kravin, D., Coates, G. D., & Carrol, M. D. (2007). *English language learners in the mathematics classroom.* Thousand Oaks, CA: Corwin Press.

College Board. (2000). The College Board: Preparing, inspiring, and connecting. (Online). Retrieved from www.collegeboard.org/prof/

Costa, A. L. (1991). *Teaching for intelligent behavior: Outstanding strategies for strengthening your students' thinking skills.* Bellevue, WA: Bureau of Education and Research.

Covey, S. (1996). *The seven habits of highly effective people.* Salt Lake City: Covey Leadership Center.

Cuevas, G. J., & Yeatts, K. (2001). *Navigating through algebra in Grades 3–5.* Reston, VA: National Council of Teachers of Mathematics.

Day, R., Kelley, P., Krussel, L., Lott, J. W., & Hirstein, J. (2001). *Navigating through geometry in Grades 9–12.* Reston, VA: National Council of Teachers of Mathematics.

DeFina, P. (2003). *The neurobiology of memory: Understand, apply, and assess student memory.* Paper presented at the Learning and the Brain Conference, Cambridge, MA.

Dewey, J. (1934). *Art as experience*. New York: Minton, Balch, & Co.

Donovan, S., & Bransford, J. (2005). *How students learn mathematics in the classroom*. Washington, DC: National Academy Press.

Ekwall, E. E., & Shanker, J. L. (1988). *Diagnosis and remediation of the disabled reader* (3rd ed.). Boston: Allyn & Bacon.

Feinstein, S. G. (2004). *Secrets of the teenage brain: Research-based strategies for reaching and teaching today's adolescents*. Thousand Oaks, CA: Corwin Press.

Findell, C. R., Small, M., Cavanagh, M., Dacey, L., Greenes, C. E., & Sheffield, L. J. (2001). *Navigating through geometry in prekindergarten–Grade 2*. Reston, VA: National Council of Teachers of Mathematics.

Friel, S., Rachlin, S., & Doyle, D. (2001). *Navigating through algebra in Grades 6–8*. Reston, VA: National Council of Teachers of Mathematics.

Gardner, H. (1983). *Frames of mind: The theory of multiple intelligences*. New York: Basic Books.

Gavin, M. K., Belkin, L. P., Spinelli, A. M., & St. Marie, J. (2001). *Navigating through geometry in Grades 3–5*. Reston, VA: National Council of Teachers of Mathematics.

Gersten, R., & Chard, D. (1999). Number sense. Rethinking arithmetic instruction for students with mathematical disabilities. *Journal of Special Education, 44,* 18–28.

Glasser, W. (1990). *The quality school: Managing students without coercion*. New York: HarperCollins.

Glasser, W. (1999). *Choice theory: A new psychology of personal freedom*. New York: HarperCollins.

Gordon, W. (1961). *Synectics*. New York: Harper & Row.

Greenes, C., Cavanagh, M., Dacey, L., Findell, C., & Small, M. (2001). *Navigating through algebra in prekindergarten–Grade 2*. Reston, VA: National Council of Teachers of Mathematics.

Gregory, G., & Chapman, C. (2002). *Differentiated instruction: One size doesn't fit all*. Thousand Oaks, CA: Corwin Press.

Gregory, G. H., & Parry, T. (2006). *Designing brain-compatible learning*. (3rd ed.). Thousand Oaks, CA: Corwin Press.

Guerrero, S., Walker, N., & Dugdale, S. (2004, Spring). Technology in support of middle-grade mathematics: What have we learned? *Journal of Computers in Mathematics and Science Teaching, 23,* 5–20.

Hannaford, C. (1995). *Smart moves: Why learning is not all in your head*. Arlington, VA: Great River Books.

Hiebert, J., Carpenter, T. P., Fennema, E., Fuson, K. C., Wearne, D., Murray, H., et al. (1997). *Making sense: Teaching and learning mathematics with understanding*. Portsmouth, NH: Heinemann.

Higbee, K. L. (1987). Process mnemonics: Principles, prospects, and problems. In M. A. McDaniel & M. Pressley (Eds.), *Imagery and related mnemonic processes: Theories, individual differences and applications* (pp. 407–427). New York: Springer.

Jackson, F. (2002). Crossing content: A strategy for students with learning disabilities. *Intervention in School and Clinic, 37*(5), 279–283.

Jensen, E. (2000). Moving with the brain in mind. *Educational Leadership, 58*(3), 34–37.

Jensen, E. (2001). *Arts with the brain in mind*. Alexandria, VA: Association for Supervision and Curriculum Development.

Jensen, E. (2005). *Top tunes for teaching: 977 song titles and practical tools for choosing the right music every time*. Thousand Oaks, CA: Corwin Press.

Jensen, E., & Dabney, M. (2000). *Learning smarter: The new science of teaching*. Thousand Oaks, CA: Corwin Press.

Johnson, D. T. (2000). *Teaching mathematics to gifted students in a mixed ability classroom*. ERIC Clearinghouse on Disabilities and Gifted Education. (EC Digest, No. E594)

Kohn, A. (1999). *The schools our children deserve: Moving beyond traditional classrooms and "tougher standards."* Boston: Houghton Mifflin.

Krepel, W. J., & Duvall, C. R. (1981). *Field trips: A guide for planning and conducting educational experiences.* Washington, DC: National Education Association.

Kuhlmann, S., Kirschbaum, C., & Wolf, O. T. (2005, March). Effects of oral cortisol treatment in healthy young women on memory retrieval of negative and neutral words. *Neurobiology of Learning and Memory, 83,* 158–162.

Lakoff, G., & Johnson, M. (1980). *Metaphors we live by.* Chicago: University of Chicago Press.

LeBoutillier, N., & Marks, D. F. (2003, February). Mental imagery and creativity: A meta-analytic review study. *British Journal of Psychology, 94,* 29–44.

Lieberman, A., & Miller, L. (2000). Teaching and teacher development: A new synthesis for a new century. In R. S. Brandt (Ed.) *Education in a new era.* Alexandria, VA: Association for Supervision and Curriculum Development.

Manolo, E., Bunnell, J. K., & Stillman, J. A. (2000). The use of process mnemonics in teaching students with mathematics learning disabilities. *Learning Disability Quarterly, 22,* 113–124.

Mann, L. (1999). Dance education: The ultimate sport. *Education Update, 41*(5), 41.

Markowitz, K., & Jensen, E. (1999). *The great memory book.* Thousand Oaks, CA: Corwin Press.

Marzano, R. J., Pickering, D. J., & Pollack, J. E. (2001). *Classroom instruction that works.* Alexandria, VA: Association for Supervision and Curriculum Development.

McCarthy, B. (1990). Using the 4-MAT system to bring learning styles to schools. *Educational Leadership, 48*(2), 31–37.

McCormick Tribune Foundation. (2004). *What every child needs* [DVD]. Chicago: Chicago Production Center.

McNamara, T. J. (2006). *Key concepts in mathematics: Strengthening standards practice in Grades 6–12* (2nd ed.). Thousand Oaks, CA: Corwin Press.

McTighe, J. (1990). *Better thinking and learning.* Baltimore: Maryland State Department of Education.

McTighe, J. (1997, October). *Performance-based instruction: Teaching and assessing for understanding.* Paper presented at the 1997 Association for Supervision and Curriculum Development Conference, Orlando, FL.

Montague, M. (1997). Student perception, mathematical problem solving, and learning disabilities. *Remedial and Special Education, 18*(1), 46–53.

Myller, R. (1991). *How big is a foot?* (Grades 1–3). New York: Yearling.

National Council of Teachers of Mathematics (Eds.). (1970). *Mathematics and humor.* Reston, VA: Author.

National Council of Teachers of Mathematics. (2006). Curriculum focal points for mathematics in prekindergarten through grade 8. (Retrieved October 2006 from http://nctm.org/standards/focalpoints.aspx?id=298.)

National Council of Teachers of Mathematics. (2008). *Response to national math panel: Factors that boost mathematical achievement.* (Retrieved 4/12/2008 from www.nctm.org.)

National Council of Teachers of Mathematics (Eds.). (2000). *Principles and standards for school mathematics.* Reston, VA: Author.

National Research Council. (2001). *Adding it up: Helping children learn mathematics.* J. Kilpatrick, J. Swafford, & B. Findell (Eds.). Mathematics Learning Study Committee, Center for Education, Division of Behavioral and Social Sciences and Education. Washington, DC: National Academy Press.

Payne, R. K. (2001). *A framework for understanding poverty.* (Rev. ed.). Highlands, TX: Aha! Process, Inc.

Posamentier, A. S., & Hauptman, H. A. (2006). *101+ great ideas for introducing key concepts in mathematics: A resource for secondary school teachers* (2nd ed.). Thousand Oaks, CA: Corwin Press.

Posamentier, A. S., & Jaye, D. (2005). *What successful math teachers do, Grades 6–12: 79 research-based strategies for the standards-based classroom.* Thousand Oaks, CA: Corwin Press.

Prystay, C. (2004, December 13). As math skills slip, U. S. schools seek answers from Asia. *The Wall Street Journal,* pp. A1–A8.

Pugalee, D. K., Frykholm, J., Johnson, A., Slovin, H., Malloy, C., & Preston, R. (2001). *Navigating through geometry in Grades 6–8.* Reston, VA: National Council of Teachers of Mathematics.

Ronis, D. L. (1999). *Brain-compatible mathematics.* Thousand Oaks, CA: Corwin Press.

Ronis, D. L. (2006). *Brain-compatible mathematics* (2nd ed.). Thousand Oaks, CA: Corwin Press.

Rothstein, A. S., Rothstein, E., & Lauber, G. (2006). *Write for mathematics* (2nd ed.). Thousand Oaks, CA: Corwin Press.

Schwartz, D. M. (2001). *On beyond a million: An amazing math journey.* Oklahoma: Dragonfly Books.

Secretary's Commission on Achieving Necessary Skills. (1991). *What work requires of schools.* A SCANS Report for America 2000. Washington DC: U.S. Department of Labor.

Solomon, P. G. (2006). *The math we need to know and do in grades PreK–5: Concepts, skills, standards, and assessments.* Thousand Oaks, CA: Corwin Press.

Sousa, D. A. (2000). *How the brain learns* (2nd ed.). Thousand Oaks, CA. Corwin Press.

Sousa, D. A. (2001). *How the special needs brain learns.* Thousand Oaks, CA: Corwin Press.

Sousa, D. A. (2006). *How the brain learns* (3rd ed.). Thousand Oaks, CA: Corwin Press.

Sousa, D. A. (2007). *How the brain learns mathematics.* Thousand Oaks: CA: Corwin Press.

Sprenger, M. (1999). *Learning and memory: The brain in action.* Alexandria, VA: Association for Supervision and Curriculum Development.

Sprenger, M. (2006a). *Becoming a "wiz" at brain-based teaching: How to make every year your best year* (2nd ed.). Thousand Oaks, CA: Corwin Press.

Sprenger, M. (2006b). *Memory 101 for educators.* Thousand Oaks, CA: Corwin Press.

Sternberg, R. J., & Grigorenko, E. L. (2000). *Teaching for successful intelligence: To increase student learning and achievement.* Thousand Oaks, CA: Corwin Press.

Storm, B. (1999). The enhanced imagination: Storytelling? Power to entrance listeners. *Storytelling, 2*(2).

Strauss, V. (2003, December 1). Trying to figure out why math is so hard for some. *The Washington Post.*

Sylwester, R. (2000). *A biological brain in a cultural classroom: Applying biological research to classroom management.* Thousand Oaks, CA: Corwin Press.

Tate, M. L. (2003). *Worksheets don't grow dendrites: 20 instructional strategies that engage the brain.* Thousand Oaks, CA: Corwin Press.

Tate, M. L. (2004). *"Sit and get" won't grow dendrites: 20 professional learning strategies that engage the adult brain.* Thousand Oaks, CA: Corwin Press.

Tate, M. L. (2005). *Reading and language arts worksheets don't grow dendrites: 20 literacy strategies that engage the brain.* Thousand Oaks, CA: Corwin Press.

Tate, M. L. (2006). *Shouting won't grow dendrites: 20 techniques for managing a brain-compatible classroom.* Thousand Oaks, CA: Corwin Press.

Tate, M. L. (2008). *Graphic organizers and other visual strategies.* Thousand Oaks, CA: Corwin Press.

Thompson, F. M. (1994). *Hands-on math! Ready-to-use games and activities for Grades 4–8.* San Francisco: Jossey-Bass.

Tileston, D. W. (2004). *Training manual for what every teacher should know.* Thousand Oaks, CA: Corwin Press.

Tompart, A. (1997). *Grandfather Tang's story.* Oklahoma: Dragonfly Books.

Wall, E. S., & Posamentier, A. S. (2006). *What successful math teachers do, Grades PreK–5: 47 research-based strategies for the standards-based classroom.* Thousand Oaks, CA: Corwin Press.

Webb, D., & Webb. T. (1990). *Accelerated learning with music.* Norcross, GA: Accelerated Learning Systems.

Westwater, A., & Wolfe, P. (2000). The brain-compatible curriculum. *Educational Leadership, 58*(3), 49–52.

Wilson, F. R. (1999). *The hand: How its use shapes the brain, language, and human culture.* New York: Vintage Books.

Willis, J. (2007, Summer). *The neuroscience of joyful education.* Retrieved July 20, 2007, from www.ascd.org.80.

Wolfe, P. (2001). *Brain matters: Translating research into classroom practice.* Alexandria, VA: Association for Supervision and Curriculum Development.

Index